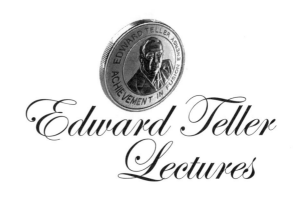

Edward Teller
Lectures

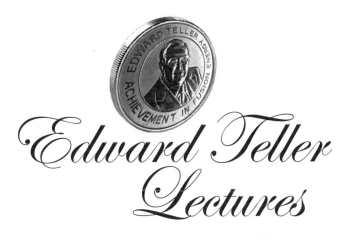

Edward Teller Lectures

Lasers and Inertial Fusion Energy

editors

Heinrich Hora

University of New South Wales, Australia

George H. Miley

University of Illinois, Urbana, USA

with a foreword by

E.M. Campbell

Imperial College Press

Published by

Imperial College Press
57 Shelton Street
Covent Garden
London WC2H 9HE

Distributed by

World Scientific Publishing Co. Pte. Ltd.
5 Toh Tuck Link, Singapore 596224
USA office: 27 Warren Street, Suite 401-402, Hackensack, NJ 07601
UK office: 57 Shelton Street, Covent Garden, London WC2H 9HE

British Library Cataloguing-in-Publication Data
A catalogue record for this book is available from the British Library.

EDWARD TELLER LECTURES: LASERS AND INERTIAL FUSION ENERGY

ISBN   1-86094-468-X

Printed in Singapore by Mainland Press

Edward Teller

(15 January 1908 – 9 September 2003)

Oil-Study by
Robin Lawrence (New York & Sydney)
After studies in Sydney, she continued
postgraduate work four years in Florence
including a work period with Oskar
Kokoschka at Salzburg and studying at St.
Martins in London. Group-shows in New
York, Camden Prize, short-listed for the
Archibald Prize

If we truly hope to live together as free people in peace, we must learn to act at a considered pace, with moderation and tolerance, and, as nations, we must provide practice in working together cooperatively for the benefit of all.

Edward Teller

(Memoirs, 2001, p. 568)

Contents

Foreword by Edward Michael Campbell
Vice-President, General Atomics, San Diego, CA

The Edward Teller medal has evolved into one of the most prestigious awards that recognizes the outstanding contributions to the field of inertial confinement fusion and high energy density science. It is appropriate that this international award be named after Teller, who with extraordinary vision and scientific insight, anticipated and then played a major role in the creation of this field.

This book - following an introduction by Professors Hora and Miley - presents the Teller winners and a sampling of their acceptance remarks. The introduction briefly summarizes the enormous and prophetic contributions of Teller and describes the history and diverse approaches to achieving ignition and gain in Inertial Confinement Fusion (ICF). It is very clear that scientific and technical innovation over the past 40 years gives confidence in the ultimate success of the ICF concept and with construction of megajoule class facilities in the United States and France, the laboratory demonstration of fusion ignition and net gain will take place within a decade. This result, analogous to Fermi's controlled fission demonstration under the squash courts in Chicago, will catalyze fusion energy development worldwide.

The Teller Award winners and their lectures provide ample evidence of the quality of researchers, and of the programs and richness of the field. They give a historical perspective and this summarizes outstanding contributions to laser-plasma coupling physics, energy transport, hydrodynamics and implosions, laser technology, numerical simulation and target design, and x-ray lasers to name only a subset.

The Teller Award Ceremony has recently become a much anticipated event at the Inertial Fusion Science and Applications Conferences. This international conference has now become the leading meeting where researchers of all the various branches and applications of Inertial Fusion come together to share results. Edward Teller would be both surprised and honored with the outstanding work and the international flavor of this meeting. He would be justifiably proud of the field of which he has fathered and of the recipients of the award named for him.

San Diego, 5 September 2003   Mike Campbell

Introductory Remarks to the "Edward Teller Lectures"

Heinrich Hora and George H. Miley

1. Overview

Edward Teller was celebrated by the media together with Albert Einstein as one of the greatest and most influential scientist of the 20th century. He had to succeed not only by his many scientific achievements but also by having the strength of character and courage to take controversial positions on the role of science and technology in the cold war. Many believe that only Teller's unique strength allowed him to understand how science could serve as a way of saving mankind from totalitarianism and oppressive dictatorships [1].

The challenge of developing controlled fusion energy by lasers and other energy sources was only a part of Teller's wide range of interests and achievements. Indeed his personnel influence and encouragement in this field was far more wide-ranging than is usually recognized. Now after 40 years of enthusiastic work on laser fusion, we can see that laser-driven inertial fusion energy (IFE) has emerged as a serious energy option. This development was initiated out of the fact that Teller in 1960, immediately envisioned after the discovery of the laser, for its use to create fusion energy. Indeed this early insight was mirrored by his ingenious counterpart Andréi D. Sakharov.

However, Sakharov's work was in a somewhat of a different direction. The carefully underlined section in Sakharov's editorial remarks for his collected papers [2] is most significant: "In a seminar in 1960 (perhaps 1961, I do not recall), I discussed the possibility of realizing a controlled thermonuclear reaction by means of lasers". Sakharov's attention to the laser was developed in his paper 1948 where he also mentioned the possibly first observation of superradiance (laser) emission by Kopfermann and Ladenburg of 1931 and the work of Fabrikant of 1937, see page 43 of Ref. [2].

Teller's familiarity with the laser precedes the work of Charles Townes who, with James Gordon, produced the very first microwave laser in 1983. In Teller's memoirs, he cites a correspondence with John von Neumann [3,p. 408]: "I suggested using Einstein's very early ideas about stimulated emission. My idea, which turned up later in some way of our correspondence, describes a laser and anticipated that invention several years. But, of course, an idea is unimportant unless one develops it." This historical remark is very important in view of the extreme difficulties Townes had when promoting his experimental discovery of the laser [4]. During this project two Nobel laureates (Rabi and Kusch) - whose "research depended on the support from the same source" - tried to stop him by saying " 'Look, you should stop the work you are doing. It isn't going to work. You know it's not going to work, we know it's not going to work. You're wasting money. Just stop!' ", see page 65 of Ref. [4]. However, three months later, the laser worked. When Townes presented this to an APS meeting, the response was not overwhelming. Apart from remarks like that of a Columbia-theoretician who flatly said that this cannot be true since it is against all rules of physics, others simply ignored the results. "Niels Bohr in Copenhagen originally said 'but this is not possible' [4]. Bohr later modified this statement more as a courtesy" when Townes told him how it worked.

After all of this turmoil it is remarkable that Townes writes [4]. "John von Neumann said in Princeton 'that cannot be right!' but few moments later understood it and wrote a famous proposal to Edward Teller". Therefore it was most fortunate that Teller ingeniously had pre-developed the scheme of the laser before in the correspondence with von Neumann who responded then so positively to Townes' experiment. This was one of the "Sternstunden der Menschheit" [5].

It was then only logical that in 1960 Teller initiated a program to study laser fusion at the Lawrence Livermore National Laboraotory. Fortunately he got his associate, John Nuckolls, involved [3]. Nuckolls had specialized in studying the smallest possible nuclear explosions and could bring this experience to bear on laser target studies. Over all these years, Teller never said that laser fusion had achieved a solution for energy production, but he was optimistic. His lecture at the IEEE conference on Quantum Electronics in Montreal [6] 1972 was a most stimulating promotion for the study of laser fusion energy [7]. Another crucial moment in 1979, Teller's cardinal merit to keep laser fusion going occurred was his intense focus and willingness to fight for the concept [3 p. 523]. In early 1979, when Greg Canavan was chief administrator of IFE in the Department of Energy there was an effort to move the IFE budget into magnetic fusion. In Canavan's fight to keep IFE alive, he called on Teller to help to prevent the cancelling of the budget. Teller spent a long afternoon in a congressional hearing "strenuously arguing for the cause" of IFE. At 71 years of age, he was then "exceedingly exhausted". The following morning his medical advisor ordered absolute resting. Then came the news about Three Miles Island and Teller forgot about being tired. Unfortunately after three days of vigorous action, he was hospitalised with a heart attack.

Thus Edward Teller was not only involved with the creation of the laser and among very first experts to initiate laser fusion energy research, but he continued to lead the field, not only with his prestige but with his intense mental and physical strength.

It was then quite natural that the executive board of the conference series "Laser Interaction and Related Plasma Phenomena LIRPP" established the "Edward Teller Medal" with the blessing by Edward Teller in 1990*. This medal is to recognize outstanding research contributions in the development of laser fusion. It is now formally incorporated as an award of the American Nuclear Society thanks to the efforts by G. Miley and by the present Board of Directors, M.E. Campbell, A. Migus, K. Mima and E. Storm and the Managing Director, W. Hogan, for the "Inertial Fusion Science and Application IFSA" conferences (the continuation of the LIRPP-series since 1999).

This book provides a collection of the lectures by the recipients of the Edward Teller Medal presented including the IFSA-Conference 8-12 September 2003 in Monterey, CA. It also presents the speeches by Edward Teller, delivered at various LIRPP conferences, over thirty years prior to when it changed into IFSA. These speeches bring out a vigorous dynamic with a mixture of various views of these developments for the creation of a key energy source for the future: inertial fusion. There is the possibility that *laser fusion energy may be generated at considerably lower costs* than any other present energy source on earth. This is exactly what mankind expected from nuclear energy when it was first discovered. This may finally lead what mankind expected: the 'Golden Age', at least with respect to energy source [8].

Why has it taken so long - more than forty years - before the optimism to offer the option of laser driven inertial fusion energy (IFE) could be substantiated? The optimistic statements by the early promoters can be traced back to underestimating the physics difficulty of compressing and heating micro-targets. The measurement of the first fusion neutrons

---

*The Advisory Board with inclusion of John Nuckolls resolved a constitution unanimously in a Board Meeting in Osaka, April 1991, to invite unrestricted nominations and by secret ballot of the Board Members – later including the awardees – through an awards committee, the awardees were all elected and in all cases approved by E. Teller.

produced by lasers was reported in 1968/69 and over the following thirty years the number of fusion neutrons per laser shot was increased by 100 billion times. Thus, the fusion gain is now close to break-even, but not yet at the level needed for a fusion power station. It is expected that the National Ignition Fusion (NIF) laser facility nearing completion at LLNL, will demonstrate fusion ignition and modest fusion gain.

A great step forward occurred during this time in laser technology. After sophisticated techniques permitted the generation of laser pulses in the femtosecond (fs) range [9,10], there was the essential discovery by Gerard Mourou of the chirped pulse amplification (CPA) of ultra short pulses [11] giving up to several Petawatt (PW) power [12] which development totally changed the field of laser applications enabling relativistic laser-plasma research [13] with numerous applications [14,15].

The Centurion-Halite experiments with underground nuclear explosions [16] clearly demonstrated experimentally that x-ray driven ICF works but at energies not available in the laboratory, needing at least some dozens MJ of driver energies [17]. Using the much more controllable energy of a laser pulse in the ns range with 10 MJ input energy a fusion yield of 1000 MJ can be expected [18]. This is an interesting definite option for laser fusion which could be studied with the newer facilities such as NIF being developed in the U.S. [19] and discussed in Section 2 and the LMJ [20]. Since the discovery [11,12] of the CPA by Mourou, the verification of PW-ps laser pulses [21-23] and the speculation, now possible, on reaching Exawatt and Zettawatt pulses, it pays to search for alternatives or modifications of the fast igniter concept [24] (Section 3). One route can be seen in the fact that laser pulses of about 10-100 kJ energy may produce a fusion energy output of 100 MJ or more if the laser pulse can be designed to ignite a fusion detonation wave. In this approach, the wave propagates in a controlled way into very large fuel mass of a low compression (solid state density or up to ten times higher) DT fuel. This process was previously considered for an electron beam ignition of targets [25] (Section 4) or for a DT ion beam block ignition [26,27] (Section 5). To emphasize this optimism the following sections detail at least some of these concepts.

2. Big Laser Solution

The present summary of the developments on laser fusion focuses on the last 10 years extending the abbreviated view given in "30 years Laser Interaction and Related Plasma Phenomena" added at the end of the here reproduced Edward Teller Lectures [28]. One key well-known earlier result [29] was the measurement of laser compression of a carbon polymer containing deuterium and tritium to 2000 times the solid-state density. The temperatures achieved, however, were disappointing low, in the range of 300 eV. As a consequence, Mike Campbell [30] explained in his contribution to a celebration of Chiyoe Yamanaka's laser research that he and colleagues, notably Max Tabak, had originally suggested additional heating should be done using short laser pulses. Campbell could develop a program to study this approach on the newly discovered CPA technique of Mourou [11,21]. This effort added new emphasis to the scheme of the fast igniter [31]. Details about this are discussed further in the following Sections. The initial aim [31] was to concentrate the ps additional laser energy into the center of the highly compressed plasma and to initiate a spark (or central core) ignition [32].

An alternative to the short pulse heating of the compressed plasma with spark ignition [32] was revealed by following up the numerical results of the volume ignition [33,34]. At this time, Atzeni [35] discussed laser fusion schemes using laser pulses of some MJ energy [19,20] in the ns range.

The most studied approach for added heating is "conventional" spark ignition where the laser compresses the plasma [7,18,32,36] in a prescribed way such that a central region core is generated at high temperatures and low density and an outer shell of high density and low temperature. The core ignites on "spark" fashion. A self-sustained fusion detonation wave is generated at the inner surface of the outer shell and that is directed into the high-density low temperature plasma.

In contrast to this spark ignition scheme and its' demanding density and temperature profiles, volume ignition [33] uses a natural adiabatic compression of the whole fuel at nearly uniform temperature and at nearly uniform density at each instant. Both parameters develop a linear velocity profile as prescribed by a self-similarity model. Unfortunately, volume compression results in a very inefficient DT burn if the target gain is less than 8. But at higher gains, alpha reheat in the fuel acts like an additional heating source. This resulting burn more than compensates radiation losses and fuel depletion, giving gains above 1000 per unit energy in the compressed plasma or above 100 overall [13,33,34]. Volume ignition compression and burn dynamics are rather "robust" (see Lackner, Colgate et al. [33]) and involve less chance for several instabilities and asymmetries than is the case for spark ignition.

Experiments using volume reaction with self-similarity dynamics led to the highest published fusion reaction gains from direct drive irradiation of spherical micro balloons [34,37,38] based on DT fusion neutron measurements. Gains of 1.5% per incident laser energy were reached, giving gains up to [34] 50% per energy in the compressed plasma core, with laser pulses in the 10-kJ range. With these small input energies, ignition was not reached but the measured data for compression density, temperature and gain fully followed the volume ignition plasma dynamic model.

These single-event [24,34,39] results should be mentioned as an alternative to the otherwise very extensively studied spark ignition [32] which details may not need to be repeated here. The difficulties are well known [34,39] and studied extensively [32]. We underline the aspects of volume ignition here only because there may be the possibility to find more simplified solutions [40].

Coming back to the result of low temperatures at high compression [29], volume ignition can provide a solution. If laser pulses of 5 MJ or more are available, volume ignition with compression to 3000 or 5000 times solid-state density results in optimum ignition temperatures as low as about 500 eV. These temperatures are, in fact, not much above the values already measured with laser pulses of about 100 times lower in energy. Detailed numerical evaluations confirm [34,39] that with few MJ laser pulses target conditions with natural self-similarity compression will arrive at a crossing point of parameters for high gain volume ignition. This approach avoids the need for additional short pulse laser input or sophisticated shaped density and temperature profiles as needed for spark ignition.

This consideration is based on the use of 5 MJ red laser input available next [19,20]. Beam smoothing would be used for the suppression of the 10-ps stochastic pulsation, avoiding the need for expensive large single crystals commonly used for conversion of the laser beam to higher frequencies [41-43]. Such beam smoothing [44] was originally introduced to reduce beam filamentation. However, this technique turned out to also reduce parametric instabilities by a factor of more than one hundred [43-45]. Instead of reducing beam filamentation, a clear disappearance of the 10-ps stochastic pulsation was observed by Labaune et al. [42,46]. Since hydrodynamic effects, absorption, parametric instabilities etc. were mostly the topic of research it may have been overlooked that smoothing techniques are much more crucial to overcome problems of instabilities [45] and of the stochastic pulsation [13,41]. Therefore, direct drive laser fusion with the fundamental red laser wavelength with appropriate smoothing may be a simplifying solution for the "big laser" option.

The spark ignition scheme [32] is recognized by the community as the basic approach for studying the fusion detonation physics. Thus, it provides a reasonable basis for near term MJ laser facilities experiments [19,20]. Although engineering type experiments and computer-assisted solutions have contributed to a growing insight into the basic processes involved, many physics issues remain unsolved. The physics is difficult since a hydrodynamic calculation cannot follow up the interpenetration of the energetic particles from the hot plasma into the cold plasma. Kinetic theory (Boltzmann equation) studies lack an adequate description of collisions, without an adequate treatment of the Boltzmann collision term. Single particle computations that typically already use one million particles are insufficient in view of the long-range Coulomb collisions. Extremely large computer capacities would be required for accurate fast multi particle simulations.

Nevertheless, for illustrative purpose, we use here the reasonable and very sophisticated evaluations to examine the computation of spark ignition a case with a laser pulse [18] of 10 MJ where instabilities and asymmetries were ignored. This is very useful for a comparison of spark and volume ignition in order to demonstrate some merits and differences between these schemes [34]. This method is also useful for comparing results obtained for the fusion detonation wave with other calculations [47].

Spark ignition [18] can result in a state of high compression as shown in Fig. 4 of [34]. This requires a precise compression process where a plasma core with average temperature of 10 keV and average density 250 n_s (n_s is the solid state DT density) is produced, surrounded from a cold and very dense outer shell with maximum density 2300 n_s. A fusion detonation wave is then ignited at the 0.158 mm radius of the compressed inner core (while the radius of the compressed outer shell is 0.323 mm). A detonation front is ignited by the inner core at the spherical interface between the core and the outer shell at about 400 n_s density. The reaction of the inner core then, in effect, corresponds to volume ignition for which the data of Fig. 5 of [34] can be used. The parameters used here have been reconstructed from the diagram of Fig. 4 of [34]. It is remarkable that they fit very well with the volume ignition of the core having E_o = 0.46 MJ deposited in the core of 1.45×10^{-5} cm^3, density of 250 n_s and 10 keV temperature. Indeed these conditions fit well with optimum volume ignition requirements. The fusion gain G is then close to 10, i.e. 4.69 MJ [5] of fusion energy is produced which impinges on the core surface producing an energy per trigger area of

$$E_{spark} = 1.62 \times 10^9 \text{ J/cm}^2. \tag{1}$$

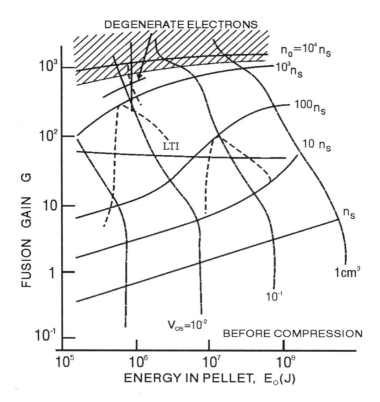

Fig. 1. Optimised core fusion gains G with the cross hatched area for Fermi degenerate electrons for evaluation of optimised volume ignition at temperatures in the 500 eV range [39].

This energy triggers the fusion detonation wave, which travels into the surrounding cold outer plasma shell.

   While one may be very skeptical about the feasibility of spark ignition due to the Rayleigh-Taylor (R-T) instabilities, a more severe problem is the spherical symmetry. (In contrast, symmetry and R-T instabilities are not crucial for volume ignition.) However, if the strongly shaped radial density and temperature profiles at spark ignition have not been properly established (cf Fig. 4 of [34]) into different radial directions, the fusion detonation wave may not at all be possible within the very short trigger time available. Apart from this problem, one may assume that the conditions for the ignition Eq.(1) had been properly scaled with respect to particle interpenetration, stopping power, radiation etc.

   It is very remarkable that the energy flux of Eq. (1) falls close to values cited by other references, (Chu, Bobin, Kidder, Bodner, Ahlborn, Babykin, and Shvartsburg [47]) where instead of (1) a value was found of

~ 10^8 J/cm^2. It was previously elaborated [47] that the inhibition of thermal conduction based on the double layer processes [13] and the collective effect of reduction of the stopping length [48,49] for the reaction products may reduce the limit (1) to about 10^7 J/cm^2. This estimate is based on possible interpenetration of the beam of energetic ions moving from the hot core plasma into the cold peripheral plasma to ignite the reaction wave. The following evaluation is based on this interpenetration model [47] but with modifications to include radiation mechanisms. As a pessimistic assumption, Brueckner and Jorna's ignition condition is used [36]. Then the ignition by irradiation of the cold plasma by fast ions requires a current density of

$$j = 10^{10} \text{ A/cm}^2. \tag{2}$$

Experts consider this early result as much too pessimistic on the hot spot spark ignition mechanism. This has to be realized when the ignition values from Eqs. (1) and (2) are employed in the following discussion of recent results from laser fusion schemes using ps laser pulses. The reference to the large laser experiments [19,20] for studying the spark ignition [32] has to be reserved for future research on the fusion detonation physics while the aspects of single-event [24] volume ignition may directly aim the application of energy production [39].

3. Physics of the Fast Igniter

Since the original paper by Max Tabak et al, on the fast igniter [31] was published, much attention and research has been focused on this unique concept. The enormous interest extended Mourou's CPA [11,12] to a new dimension involving laser fusion. Today, lasers with ps pulses of PW power are available [21,50]. The numerical evaluation of the fast igniter highlighted that estimated fusion gains, based on fusion energy per laser energy, reach values close to 300 for laser energies below MJ. Preceding schemes had typically predicted gains around 100 and required much higher energy and longer (ns) laser pulses. Nevertheless longer laser pulses, such as employed in other schemes will also be necessary for the fast igniter, perhaps with 10 times lower energy for the first step of the compression to densities above 2000 times solid state density.

    In the second step of fast ignition short PW laser pulses are deposited into the precompressed plasma. These new ideas [31] require moving into a new field of relativistic effects during such laser plasma interactions.

The initial dream guiding the PW pulse to the center of the high-density fuel with the energy of the PW pulse deposited within a very small central spark volume was fascinating but very questionable. First experiments [51] with ps pulses of powers in the 10 TW range showed the enormous complexity of the relativistic interaction. An incomplete list of interaction phenomena includes stopping lengths, electron and ion acceleration, double layer effects (now called "Coulomb acceleration"), and generation or self-focusing for guiding of the PW beam to the center etc. These phenomena were considered, though with very simplified and incomplete models. Conditions were found such that most of the PW energy may be uniformly spread over the compressed fuel for volume ignition. If this can be achieved experimentally, fast ignition [52] may arrive at interesting fusion gains.

A funnel concept was implemented experimentally to study fast ignition. This technique uses a gold cone inserted in the target [23,50] to guide the beam into the center core with complicated interactions in the outer volume plasma. The output from such experiments has increased from 10^4 DT fusion neutrons to 10^7 fusion neutrons using a cone [50].

Other recent studies with ps laser pulses of more than 3 TW power indicate a number of very unexpected phenomena can occur with these ultra high power interactions. Gammas up to 50 MeV energy appeared and subsequent nuclear photoeffect interactions produced a variety of radioactive isotopes [53-56]. These gamma bursts had intensities far above the limits of radiation safety requiring new safeguards [56]. There are also speculations about Zettwatt pulses, (see Mourou and Tajima [21]), for producing pair production in vacuum or Hawking radiation similar to black holes [57,58]. For other experiments 30 TW pulses focused into ~ atmospheric pressure gas produced 10^8 electrons conical directed in beams of 30 MeV energy [59]. These beams could be explained theoretically (in competition to other theories [60]) in number, energy and conical angle by electron acceleration by lasers in vacuum [61,62]. That process is identical to the "free wave acceleration" derived by Hartemann, Woodward et al. [63]. Experiments by Malka et al. [64] measuring 200 MeV electrons at similar conditions as of Umstadter et al. [59] can quantitatively be explained by the free wave acceleration [60,63] in contrast to vaguely assumed wake field models. There appears to be some interesting application of these energetic electron beams for example, if an electron beam is used, lead target can generate positrons with an extremely high intensity by the Bhabha process [65].

4. Fast Igniter for Electron Beam Fusion

While the funnel scheme [50] is a straight forward advancement of the initial fast igniter concept [31], further generalizations evolved for the fast ignition. The nonlinear and relativistic properties of compressed plasma produced by irradiation with PW-ps laser pulses, especially the dominating conversion into extremely intense relativistic electron beams, is generally viewed as the main energy transfer process [25] involved in the fast igniter scheme [31]. In order to reach condition (1) for ignition of a detonation wave, the 10 to 100 kJ in a ps laser pulse must be focused to ~ 0.03-mm diameter. The basic issue then is if this energy can be deposited within a sufficiently short depth in the fusion fuel. When correctly fitting the observations of relativistic electrons of more than 5 MeV energy (30 TW pulses produced 30 MeV electrons in less than solid state densities [59]) the necessary short stopping length can only be reached at high plasma densities of $> 1000n_s$. This result sets the goal for precompression of the plasma in the fast igniter scheme [31].

Under these conditions, a fusion detonation wave may be ignited. Nuckolls and Wood [25] consider propagation of this wave into low density DT fuel and explain the advantages of this technique, even for densities of ten times the solid state (3 g/cm^3). It is important to note that then 10-kJ laser pulses can produce 100 MJ or more output energy, depending only on the amount of low-density fuel into where the detonation wave is propagating.

These considerations are consistent with the preceding discussions. For example, spark ignition [18] can be evaluated with the reaction of the inner core of average density ~ $300n_s$ and radius ~0.158 mm where the incorporated 0.5 MJ laser energy produces 5 MJ in full agreement with the volume ignition calculation [34,39] and the more general analysis by Atzeni [35]. This energy corresponds to 3.3×10^{10} J/g in good agreement with the basic 4×10^{10} J/g in the Nuckolls-Wood model [25]. While this agreement is encouraging, this represents only a part of the physics of fast ignition. Various researches cited here have identified a large number of effects that still need to be explored before the numerous assumptions in this spark ignition concept can be fully verified.

One important problem is to understand how the 10^8 30-MeV electrons produced by 30 TW-0.4 ps laser pulses fits so well with the free wave acceleration mechanism [60,62,63]. Collective effects for ions slowing down in such plasma [48,49] may well occur reducing the alpha stopping length such that the estimation of Nuckolls and Wood [25] in Eq. (1) could be modified to predict even more favorable conditions.

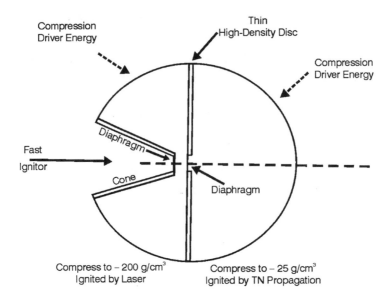

Fig. 2 Laser fusion ignition scheme of Nuckolls and Wood [25]. The PW-ps laser pulse in the pre-compressed DT plasma in the left hand half sphere produces an intense electron beam which ignites a fusion detonation wave in the right hand side half sphere with nearly uncompressed large amount of DT fuel.

5. Block Igniter for Producing Ion Beam Ignition

A further example should be mentioned here to show the richness of ICF research as a single-event [24] further generalization of the fast igniter. One aim is to provide conditions how the petawatt-picosecond laser-plasma interaction may provide a further access to study the physics of fusion reaction fronts similar to the Nuckolls-Wood scheme [25] discussed in the preceding paragraph. The following modification of the fast igniter scheme [31] using a "block" igniter is intended to avoid the high precompression of the plasma such that all the advantages of the low density, large mass targets envisioned in the Nuckolls-Wood scheme would result [25].

The experiments described next open the way to produce very intense DT ion-beam blocks for the ignition near conditions at or below the relativistic beam threshold. Such blocks avoid the various undesirable relativistic beam effects in the irradiated plasma. These blocks follow immediately from the PW-ps interaction of the irradiated solid DT fuel. It is expected that this technique will produce a reaction front in a long

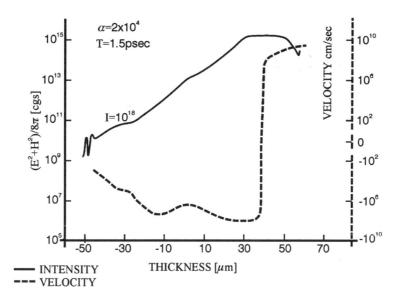

Fig. 3. Deuterium ion velocity at 1.5 ps showing the generation of blocks of deuterium plasma moving against the neodymium glass laser light (positive velocities v to the right) and moving into the plasma interior (negative velocities) at irradiation by a neodymium glass laser of 10^{18} W/cm^2 intensity onto an initially 100 eV hot and 100 μm thick bi-Rayleigh profile (Fig. 10.17 of [13]) with minimum internal reflection. The electromagnetic energy density $(\mathbf{E}^2+\mathbf{H}^2)/(8\pi)$ is shown at the same time of 1.5 ps after begin of the constant irradiation [13].

stretched target or in a half sphere [25] that fulfils the condition of a low density, large mass reaction volume. If so, 10 kJ laser pulses may well produce 0.1 to few GJ of fusion energy.

The block ignition scheme follows from the analysis of the anomalous ion emission from solid targets during ps-TW laser irradiation [15,66]. While laser pulses of about ns duration produce maximum MeV ion energies in agreement with relativistic self-focusing, ps pulses result in more than 50 times lower maximum energies. Furthermore, the numbers of emitted ions do not vary much with laser power. This experiment had a suppression of the prepulse by a factor of 10^8 (contrast ratio) until 100 ps before the main ps pulse. Thus, it was concluded [67] that relativistic self-focusing was avoided and that a pure skin-depth interaction occurs within a constant interaction volume. As theoretically expected, the measured ion energy is then proportional to the input power P. The experimentally determined quiver energy corresponded to the ion energy assuming a

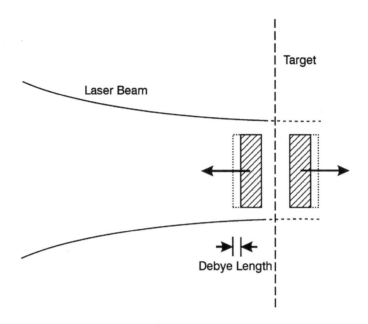

Fig. 4. Scheme of skin depth laser interaction where the non-linear force accelerates a plasma block against the laser light and another block towards the target interior. In front of the blocks are electron clouds of the thickness of the effective Debye lengths of less than 500 nm [26].

plasma-dielectric swelling of the nonlinear force by a factor of three due to the 100 ps prepulse. This conclusion is based on a contrast ratio of 10^4 during the last 100 ps before the main pulse arrives, in agreement with a numerical simulation using an extensive two-fluid calculation (Fig. 1 of Ref. [27]). This analysis of prepulse mechanisms and the contrast between skin depth vs. relativistic self-focusing interaction was evident [27] also in similar measurements with precisely controlled prepulses [68].

The generation of fast moving high density plasma blocks from direct electrodynamic forces at laser-plasma interaction is based on the complete derivation of the nonlinear force [13]. Detailed numerically studies of plane inhomogeneous plasma with perpendicular incidence of lasers [13] consider a neodymium glass laser pulse of 10^{18} W/cm^2 irradiating a 100 wave length thick deuterium plasma of up to the critical density (Fig. 3). After 1.5 ps a profile of energy density and as negative gradient produces forces that drive the plasma into blocks with velocities up to 10^9 cm/s. One block moves against the laser light while the other moves into the plasma interior. For the conditions leading to the skin layer interaction [67], this results in the block generation (Fig. 4). The nonlinear forces drive an electron cloud, such that the ions are dragged within a double layer given by the Debye length which value is sufficiently small for the conditions discussed here for laser fusion (Fig. 5).

The plasma block moving into the plasma interior can be interpreted in terms of prior studies of light-ion beam fusion. The conditions for ignition of a fusion reaction front moving into an uncompressed large volume of fusion fuel have been studied by a number of authors [47], and they find an energy flux of Eq. (1) necessary. These numbers are in fair agreement with the generation of the fusion detonation wave at spark ignition of laser fusion [18] and also with the electron-beam ignition requirement now introduced for the PW-ps laser pulses [25]. The rather pessimistic condition for initiating the fusion reaction wave of Eq. (2) following Brueckner and Jorna [36] can easily be achieved by the PW-ps laser pulses using the skin layer conditions with carefully selected prepulses to avoid relativistic self-focusing but allowing generation of some swelling.

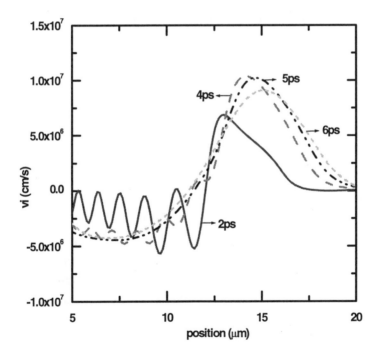

Fig. 5. Ion velocity profiles at the times 2, 4, 5 and 6 ps at irradiation of a 4 ps neodymium glass laser pulse on a deuterium plasma of initial density of a linearly increasing ramp of 30 eV temperature, confirming the generation of an ablating plasma block (negative velocity) and a compressing plasma block (positive velocity) [Y. Cang et al., J. Plasma Physics 2004, in print].

This requires the DT ion density of the compressing block is 10^{21} cm^{-3} for neodymium glass lasers. The laser intensity has to be selected in the range of 10^{18} W/cm^2 or less adjusted, such that swelling can produce ion energies of 80 keV. The resulting current density of 4×10^{10} A/cm^2 is then sufficient to fulfill even the pessimistic threshold of Eq. (2). The thickness of the igniting block can be adjusted by the swelling.

In summary, to simulate high gain laser fusion by PW-ps laser pulses is expected based on light-ion beam type ignition of the skin layer process [27,67] and associated effects measured by Badziak et al [15,66]. This is an extension of the fast ignition [31] scheme which is an alternative to Nuckolls-Wood electron-beam type fusion scheme [25]. The block ignition scheme provides a continuation of the laser fusion gain curve starting from the high gain values [69] achieved by irradiating 6 J-ps laser pulses focused on 10-μm diameter deuterium targets.

The prepulse controlled nonlinear-force driven skin layer block igniter may be extended later to use p^{11}B fuel since ever present PW laser intensities can produce ion energies in the blocks up to MeV or more. Thus energies required of the block ions for p^{11}B ignition, (e.g. to go to several 100 keV energy to match the resonance of the reaction) appears to be achievable. In this case, the plane wave skin layer interaction could even employ lower laser intensities (around or even less than the relativistic threshold) avoiding the numerous relativistic anomalies. If achieved p^{11}B fusion plants could use direct conversion of the reaction particle energies into electricity [70] less radioactivity per gained electricity is produced than from coal plants [13]. Such plant would also involve a minimum of waste heat generation, reducing the huge cooling burden of today's power stations.

There are well a number of problems to be solved for the block igniter, but the basic mechanism of the skin-layer laser-plasma interaction at sub-relativistic conditions in plane geometry without relativistic self-focusing has been confirmed experimentally [15,27,66,67]. The earlier results from light-ion beam fusion [47] provide insight into the conditions needed for driving reaction waves into large volumes of solid-state density DT fuel. If successful, this would give extremely high gains avoiding the need for complicated precompression processes.

6. Concluding Remarks

Since this book collects the results of many very outstanding experts in the field of laser fusion, why were these remarks about possible ignition schemes given as an introduction? The most imperative reason was that this provided an opportunity to give references to lesser known documents about how the laser was created and how Edward Teller contributed to it's early development. Further his essential support for laser fusion had to be underlined. To illustrate how Teller's statement that laser fusion may be not too far away could be implemented, recent results with the large scale NIF-like laser for a technologically clarified solution for energy production and new aspects using PW-ps laser pulses have been developed in some detail. The numerous different points in the following Edward-Teller-Lectures underline further what a wide range of science must be considered in search for useful laser fusion. Still we are only at a beginning, but clearly aimed toward the most important goal envisioned by Teller of very low cost, safe, clean and in exhaustive energy generation. The way to this goal seems clear now that the main routes, either with the ns or with the ps pulses - or both, have identified.

References:

1. Laser and Particle Beams 19, 665 (2001)..
2. A.D. Sakharov, *Collected Scientific Works*, Marcel Dekker, New York and Basel 1983, see Laser and Particle Beams 5, 163 (1987)
3. Edward Teller, *Memoirs. A Twenty-Century Journey in Science and Politics*, Perseus Publishing, Cambridge Mass. 2001, see [1].
4. Charles Townes, *How the Laser Happened*, Oxford University Press, New York, 1999, see Laser and Particle Beams 18, 151 (2000)
5. "Starmoments of mankind" as the German-Austrian literature-Nobel-laureate Stefan Zweig said.
6. E. Teller, IEEE J. Quantum Electronics 8, 564 (1972); Bull. Am. Phys. Soc. 17, 1034 (1972)
7. J.H. Nuckolls, *Laser Interaction and Related Plasma Phenomena*, H. Schwarz and H. Hora eds. (Plenum, New York 1974) Vol. 3B, p. 399
8. S. Eliezer and H. Hora, Direct driven Laser Fusion, in *Nuclear Fusion by Inertial Confinement*, G. Velarde, J. Martinez-Val and A. Ronen eds. (CRC Press, 1993) p.43-72
9. C.V. Shank, R. Yen and C. Hirlimann, Phys. Rev. Letters 50, 434 (1983)
10. J.-C. Diels, W. Dietel, J.J. Fontaine, W. Rudolph and B. Wilhelmi, J. Opt. Soc. Am., B2, 680 (1985)
11. D. Strickland and G. Mourou, Optics Communications 56, 219 (1985)
12. G. Mourou and R. Umstadter, Scientific American 286 No.5, 81 (2002)

13. H. Hora, *Plasmas at High Temperature and Density*, Springer, Heidelberg 1991 (paperback: S. Roderer, Regensburg/Germany 2000)
14. E.L. Clark, K. Krushelnick, M. Zepf, F.N. Beg, M. Tatarakis, A. Machacek, M.I.K. Santala, I. Watts, P.A. Norreys, and A.E. Dangor, Phys. Rev. Letters, 85, 1654 (2000)
15. J. Badziak, H. Hora, E. Woryna, S. Lablonski, L. Laska, P. Parys, K. Rohlena, and J. Wolowski, Phys. Letters A315, 452 (2003)
16. W.J. Broad, New York Times 137 No. 47451 March 12 (1988)
17. C.R. Phipps jr., Laser and Particle Beams 7, 835 (1989)
18. E. Storm, J. Lindl, E.M. Campbell, T.P. Bernat, L.W. Coleman, J.L. Emmett, et al., Lawrence Livermore National Laboratory, Report No. 47312 (1988)
19. C.B. Tarter, *Inertial Fusion Science and Applications 2001*, K.A. Tanaka, D.D. Meyerhofer and J. Meyer-ter-Vehn eds., (Elsevier, Paris 2002) p. 9
20. Rene Pellat, *Inertial Fusion Science and Applications 2001*, K.A. Tanaka, D.D. Meyerhofer and J. Meyer-ter-Vehn eds., (Elsevier, Paris 2002) p. 17
21. M.D. Perry and G. Mourou, Science 264, 917 (1994); G. Mourou and T. Tashima, *Inertial Fusion Science and Applications 2001*, K.A. Tanaka, D.D. Meyerhofer and J. Meyer-ter-Vehn eds., (Elsevier, Paris 2002) p. 831,
22. T.E. Cowan, M.D. Perry, M.H. Key, T.R. Ditmire, S.P. Hatchett, E.A. Henry, J.D. Moody, M.J. Moran, D.M. Pennington, T.W. Phillips, T.C. Sangster, J.A. Sefcik, M.S. Singh, R.A. Snavely, M.A. Stoyer, S.C. Wilks, P.E. Young, Y. Takahashi, B. Dong, W. Fountain, T. Parnell, J. Johnsn, A.W. Hunt, and T. Kühl Laser and Particle Beams 17, 773, (1999)
23. R. Kadama et al., Nature 412, 789 (2001)
24. M.H. Key, Nature 412, 775 (2001)
25. J.L Nuckolls, and L. Wood, *Future of Inertial Fusion Energy*, Preprint UCRL-JC-149860 (September 4, 2002) www.llnl.gov/tid/Library.html
26. H. Hora, F. Osman,R. Höpfl, J. Badziak, P. Parys, J. Wolowski, E. Woryna, F. Boody, K. Jungwirth, B. Kralikova, J. Kraska, L. Laska, M. Pfeifer, K. Rohlena, J. Skala, J. Ullschmied, Czechoslov. J. Physics 52, Suppl D as CD-ROM in Issue No. 7, D349 (2002); H. Hora, Czechoslov. J. Physics 53 (2003)
27. H. Hora, Hansheng Peng, Weiyan Zhang, and F. Osman, *High Power Lasers and Applications II, SPIE Proceedngs*, Vol. 4914, Dianyuan Fan, Keith A. Truesdell, and Koji Yasui eds., p. 37-48 (2002); H.

Hora, J. Baziak et al., *European Conference on Laser Interaction with Matter, SPIE Proceedings,* Vol. 5228, p.253 (2003)

28. H. Hora, *Laser Interaction and Related Plasma Phenomena,* G.H Miley and E.M. Campbell eds., AIP Conference Proceedings No. 406, Am. Inst. Phys. Woodbury, NY, 1997, p. XV, this book p.

29. H. Azechi, et al. 1991 Laser and Particle Beams 9, 167

30. E.M. Campbell et al, *Light and Shade: Festschrift to the 77th Birthday of Chiyoe Yamanaka,* Osaka University 2000, p. 471

31. M. Tabak et al., Physics of Plasmas 1, 1626 (1994)

32. J.D. Lindl, Physics of Plasma 2, 3933 (1995)

33. H. Hora and P.S. Ray, Zeitschr. f. Naturforschung A33, 890 (1978); R.C. Kirkpatrick and J.A. Wheeler, Nucl. Fusion 21, 389 (1981); R.J. Stening, R. Khoda-Bakhsh, P.Pieruschka, G. Kasotakis, E. Kuhn, G.H. Miley and H. Hora, *Laser Interaction and Related Plasma Phenomena,* G.H. Miley et al ed., (Plenum New York 1991) Vol. 10, p. 347; R. Khoda-Bakhsh, Nuc. Instr. Meth. A330, 263 (1993), X.T. He and Y.S. Li, *Laser Interaction and Related Plasma Phenomena.*G.H. Miley ed.AIP Conf Proceedings No. 318 (Am. Inst.Physics, New York 1994) p. 334; K.S. Lackner, S.A. Colgate, N.I. Johnson, R. Kirkpatrick, and A.G. Petschek, *Laser Interaction and Related Plasma Phenomena,* G.H. Miley ed., AIP Conf. Preceed. No. 318 (Am. Inst. Phys., New York 1994) p. 356; J.-M- Martinez-Val, S. Eliezer and M. Piera, Laser and Particle Beams 12, 681 (1994); N.A. Tahir and D.H.H. Hoffmann, Fusion Engin. and Design 24, 418 (1994); S. Atzeni, Jap. J. Physics 34, 1986 (1995)

34. H. Hora, H. Azechi, Y. Kitagawa, K. Mima, M. Murakami, S. Nakai, K. Nishihara, H. Takabe, C. Yamanaka, M. Yamanaka, and T. Yamanaka, J. Plasma Physics, 60, 743 (1998)

35. S. Atzeni, *Inertial Fusion Science and Applications 2001,* K.A. Tanaka, D.D. Meyerhofer and J. Meyer-ter-Vehn eds., (Elsevier, Paris 2002), p. 45

36. K.A. Brueckner and S. Jorna, Rev. Mod. Phys., 46, 325 (1974)

37. H. Takabe et al. Phys. Fluids 31, 2884 (1988)

38. J.M. Soures, R.L. MacCrocy et al., Phys. Plasmas 3, 2108 (1996)

39. H. Hora, G.H. Miley, and F. Osman, J. Plasma Physics, 69, 413 (2003)

40. Peter Amendt, H.-S. Park, H.F. Robey, R.E. Tipton, R.E. Truner, J.D. Colvin, M.J. Edwards, S.W. Haan, B.A. Hammel, O.L. Landen, K.O. Mikaelian, L.L. Milovich, D.P. Rowlley, and L.J. Suter, IFSA03 Abstract Book, p. 39

41. H. Hora, and M. Aydin, Phys. Rev. A45, 6123 (1992)

42. H. Hora, and M. Aydin, Laser and Particle Beams 17, 209 (1999)
43. H. Hora, *Current Trends in Internatinal Fusion Research, Proceeings ofthe third Symposium*, E. Panaralla ed., NRC Research Press, National Research Council of Canada 2003, p. 433
44. Y. Kato et al., Phys. Rev. Letters 53, 1057 (1984); R.H. Lehmberg and S.P. Obenschain, Opt. Commun. 46, 27 (1983)
45. A. Giulietti, et al. *Laser Interaction and Related Plasma Phenomena*, H. Hora and G.H. Miley eds., Plenum , New York, Vol. 9,p. 261
46. Chr. Labaune et al., Phys. Fluids B4, 2224 (1992)
47. H. Hora, Atomkernenergie/Kerntechnik, 42, 7 (1983)
48. D. Gabor, Proc. Roy Soc. London, A213, 73 (1953)
49. P.S. Ray et al. Zeitschr. f. Naturforsch.28, 155 (1976)
50. R. Kadama & Fast-Igniter Team, Nature 418, 933 (2002)
51. A.P. Fews, P.A. Norreys et al. Plasma Physics and Controlled Fusion 73, 1801 (1994)
52. H. Hora, H. Azechi, S. Eliezer, Y. Kitagawa, J.-M. Martinez-Val, K. Mima, M. Murakami, K. Nishihara, M. Piera, H. Takabe, M. Yamanaka, and T. Yamanaka, *Laser Interaction and Related Plasma Phenomena*, G.H. Miley and E.M. Campbell eds., AIP conf. Proceedings No. 406, Am. Inst. Phys., New York 1997), p. 156
53. K.W.D. Ledingham, et al, Phys. Rev. Letters 84, 899 (2000)
54. H. Schwörer, P. Gibbon, S. Düsterer, K. Behrens, C. Ziener, C. Reich, and R. Sauerbrey, Phys. Rev. Letters 86, 2317 (2001)
55. M. Roth, M.H. Key et al., Phys. Rev. Letters 86, 436 (2001)
56. W.P. Leemans et al, Physics of Plasmas 8, 2510 (2002)
57. H. Hora, R. Castillo, W.-K. Chan, M. Collins, and T. Stait-Gardner, Eurpean Conference on Laser Interaction with Matter, Prague 2000, M. Kalal, R. Rohlena and M. Sinor eds., SPIE Conference Proceedings No. 4424, (2001) 528-532
58. H. Hora, F. Osman, R. Castillo, M. Collins, T. Stait-Gardner, Wai-Kim Chan, M. Hölss, W. Scheid, Jia-Xiang Wang, and Yu-Kun Ho, Laser and Particle Beams 20, 79 (2001)
59. R. Umstadter, Laser Focus 22 (No. 2) 101 (1996); R. Umstadter et al Sciences 273, 472 (1996)
60. H. Hora, M. Hoelss, W. Scheid, J.X. Wang, Y.K. Ho, F. Osman and R. Castillo, Laser and Particle Beams, 18, 135 (2000)
61. H. Hora, Nature 333, 339 (1988)
62. T. Häuser, W. Scheid and H. Hora, 1994, Phys. Letters A186, 189 (1994)
63. F.V. Hartemann et al., 1998, Phys. Rev. E68, 5001 (1998)

64. Rev. Letters 80, 1351 (1998), Lefebvre et al. Phys. Rev. Letters 80, 1352 V. Malka et al., Science 298, 1596 (2002); P. Mora and B. Quesnel, Phys. (1998)
65. C. Gahn, K. Witte et al, Appl. Phys. Lettrs 77, 2662 (2000)
66. J. Badziak et al. Laser and Particle Beams17, 323 (1999)
67. H. Hora, J. Badziak et al, Opt. Commun. 207, 333 (2002)
67. P. Zhang et al, Phys. Rev. E57, 3746 (1998)
69. P.A. Norreys et al. Plasma Physics and Controlled Fusion 40, 175 (1998)
70. G.H. Miley, Fusion Energy Conversion, American Nuclear Society Press, 1976; H. Hora, German Patent 10208515 A1 application 28 Feb. 2002, declassified 5 Sept. 2002; Disclosure 16 Oct. 2003; W.B. Boreham, H. Hora, M. Aydin, S. Eliezer, M.P. Goldsworthy, Gu Min, A.K. Ghatak, P. Lalousis, R.J. Stening, H. Szichman, B. Luther-Davies, K.G.H. Baldwin, R.A.M. Maddever, A.V. Rode, Laser and Particle Beams 15, 277 (1997)

Occasional Addresses by Edward Teller at Conferences of Laser Interaction and Related Plasma Phenomena (LIRPP)

Edward Teller
Troy, New York 1973

FUTUROLOGY OF HIGH INTENSITY LASERS*

Edward Teller

Lawrence Livermore Laboratories
University of California
Livermore, California   94550

I promised to talk about the futurology of high intensity lasers. That was an unwise decision on my part. I find that every talk that has been given here is on futurology, whether it is admitted or not.

I would like to discuss one of the simple questions connected with the action of very high intensity lasers because that will give me a way to illustrate how great these intensities are. An interesting point that one can get in the scale of intensities is when the light can do much more than being absorbed just once. The question is whether the light can be absorbed again and again. You absorb it once and from that point you start the second absorption and the question whether this will or will not happen is connected with one simple ratio. The numerator of this ratio is the electric field times the dipole moment that is connected with the transition. The denominator of the ratio is the energy of the transition.

This ratio gives one a measure whether few processes or many processes will occur. When this ratio becomes one then any number of processes can occur. Now the point where this ratio becomes one is when the electric field in the light wave becomes as strong as the electric field that holds the electron in an outer orbit of an atom or molecule. This is the kind of intensity that has been reached and surpassed in the focus of the lasers which we try to use to obtain laser fusion. In other words, we have the kind of

---

*Keynote Address delivered at the Third Workshop on "Laser Interaction and Related Plasma Phenomena" held at Rensselaer Polytechnic Institute, Troy, New York, August 13-17, 1973.

intensity in which the electric field is sufficient in one period
of vibration, in fact, in $1/2\pi$ periods of vibration, to tear an
electron out. Therefore, an instant plasma is produced. Such an
intensity has been reached and in some cases surpassed.

The intensities that I am talking about are high enough so
that the energies contained in the electromagnetic wave within the
volume of an atom are as great, approximately, as the ionization
energy of the atom. (This happens to be essentially the same
statement as given above.) Now, that means that within the wavelength cube we have many quanta; namely, wavelength over the atom
diameter to the 3rd power. That means for ultra-violet light which
is proper for the process of ionization approximately a million
quanta per wavelength. In the infrared (which is insufficient to
ionize) this would mean roughly a thousand billion quanta, a tril-
lion quanta per wavelength.

Now, I would like to draw your attention to a process that at
such high intensities could be relevant. I am interested in the
excitation of a vibration in a molecule; for instance, in a dia-
tomic molecule. I would like to raise the question: what light
intensity does one need so that in one period of vibration you can
excite one quantum of the vibration? At the relevant intensities
of laser light one can forget about the quantum nature of light
and one can talk about light as though it were a classical process.
For the excitation of a quantum of vibrations in one period it turns
out that we need less intensity than what we need to tear an atom
apart, to ionize an atom by the ratio of the energy of the quantum
divided by the ionization energy. It is easy to verify that this
ratio is approximately the square root of the ratio between the
electron mass and the reduced mass of the vibrating molecule. All
this, of course, is true only if there is a strong charge, i.e. a
strong change of dipole moment connected with the vibration.

This leads to an amusing possibility: You know all kinds of
people are talking about all kinds of schemes to separate isotopes
with the help of lasers. I would like to propose a new method;
one, which is in practice probably more difficult than the methods
proposed, but one which is conceptually simple. I want to absorb
enough energy to excite a vibrational quantum during one vibration
and then keep it up. In other words, I want to excite the vibration
essentially in the same way as you treat a child on a swing. You
just push at the proper time. This means that we must keep track
of how much energy has gone into the vibration, and as the vibration increases, the frequency usually decreases. We can change the
frequency by tuning the laser with the help of a magnetic field,
a method that is well known. Question: can we tune in such a way
that we do not tear the molecule apart in one vibrational period,
in other words, can we stay below the intensity needed to ionize?

FUTUROLOGY OF HIGH INTENSITY LASERS　　　　5

By using the right frequency a lower intensity will be sufficient, essentially lower by the square root of the mass ratio of the electron to the nuclear mass.

The actual experiment won't be easy. As you excite the vibration the frequency of the laser must keep pace with the energy fed into the molecule. This energy is connected with the frequency, with the effective charge, and with the amplitude. For instance, the energy that is needed so that you absorb a quantum per vibration goes down more or less proportionally to the number of excitations, but the needed intensity also changes because as the molecule vibrates, the effective vibrating charge is changing. This means that we have to keep track of an additional variable of which we don't know as much as we should.

For example, hydrogen can be torn out from hydrogen fluoride in the number of vibrations equal to the number of vibrational levels of this molecule. This, of course, would be done in a much shorter time than the number of collisions in a gas. The DF frequency is sufficiently different from the HF frequency so that there can be no mix-up between the two and you can tear apart the molecules containing deuterium, and not those containing hydrogen. This may lead to an effective isotope separation, undisturbed by molecular collisions.

If you excite a molecule like HF you can get vibrational distributions where high vibrational quanta are already present in considerable numbers. This can be done by well-known methods. When you start from high vibrational quanta, then the intensity that you have to use will become less. Actually, the beauty of the process is that if you absorb one quantum per vibration, then, effectively, you broaden each quantum so that the discrete nature of the quanta is forgotten. The molecule really behaves like a swing and you can treat the whole process (the vibrations as well as the laser light) according to classical mechanics. Incidentally, the cross sections that you get in all these processes is one per cent or somewhat less of molecular dimensions. This means that rather complete absorption of the laser light is possible.

I would like to add just a few more thoughts on the application of high intensity lasers. Yesterday I saw at AVCO a very big CO_2 laser. I was told that such intensities have been reached, that it will be possible to reflect laser light from the moon without a corner reflector. You know, laser light has been returned from the moon because corner reflectors have been dropped on the moon which have the property of throwing back the light precisely where it came from. That, of course, is a great advantage and in this way the distance to the moon has been measured to the accuracy of one foot. However, with this big laser, you can't do it to the accuracy

of a foot because the illuminated portion of the moon will be
approximately three miles across, over which there will be a variation of distance considerably more than a foot. However, by
measuring the time light went and came you can get a nice mapping
of altitudes on the moon without sending anybody there.

You can go farther. This cannot be done yet. But I am sure
that in a few years light can be reflected from the closest planets,
Mars and Venus. And taking into account curvature and everything
else, you can get the distance to Mars to an accuracy of one part
in 10^{10}. This is considerably better than present accuracies. We
might do much more on Venus, which has an atmosphere through which
we can't see. By using light of various frequencies we might be
able to penetrate to various depths into that atmosphere and the
exploration of the Venus atmosphere might get quite a considerable
impetus by reflected laser light in which you vary the frequency.

The head of the AVCO-Everett Laboratory, Arthur Kantrowitz,
a Ph.D. student of mine, has said that in the not too distant
future we will be able to send a satellite into orbit with the
help of laser propulsion by evaporating the tail end of the
satellite with big lasers. The lasers have to be installed above
most of the clouds. We have a space station in Hawaii, in Haleakala
on Maui at an altitude of about 11,000 feet where he wants to put
his laser. All he needs is a 1,000 MW laser or, preferably, ten
100 MW lasers. Now then I will make a rather unpopular statement
about that; I will say it will happen before laser fusion will make
a contribution in a practical sense. I am interested in the
question how soon the fusion energy that we want to squeeze out of
these microexplosions will really give economic power. And I
believe that propulsion of manned satellites will occur before that
occurs.

This brings me to the main topic, laser fusion. All of us
know that we should be able to cause thermonuclear explosions in
very small droplets of deuterium if only we could compress them
ten thousand-fold (perhaps even a thousand-fold will suffice).
This is quite a task. Not only because of the required laser
intensities, but also because the compression must go on in quite
a symmetrical fashion. About the latter point, one can be optimistic or pessimistic. I want to mention the definition of an expert
given by Niels Bohr: "An expert is a person who through his own
painful experience has found out about all the mistakes one can
commit in a very narrow field." I claim that in laser fusion,
there do not exist any experts yet. Most of the mistakes are ahead
of us. This does not mean that we shouldn't do it. On the contrary,
I think it is an extremely interesting field. But I think that when
you have a long way to go we should pick flowers on the way.

FUTUROLOGY OF HIGH INTENSITY LASERS   7

Coming over to R.P.I., I heard that there are experiments in
Garching in Germany which show a compression of a solid four or
five fold. This is a good beginning. Now one beautiful thing
about the laser method is, that while you compress very small objects
these objects are out in the open. The whole compression takes the
order of a billionth of a second or less. But, lasers can produce
radiation (not necessarily coherent radiation) of all kinds of wavelengths, in the ultraviolet, in the x-ray region. Lasers can
stimulate this radiation with high intensity over exceedingly short
periods. Therefore, if we compress such a small droplet we can
look through it, take an x-ray photo, try to use the wavelength which
is just able to penetrate through this little droplet and see how
the compression proceeds. Does it proceed symmetrically? Can we
see the shell of the shock wave going in? Can we find out, not
only what the average density is but what the density distribution
is?

I said that lasers can tell us a lot about astronomy, I am now
telling you that lasers could tell us a lot about the equation of
state of all kinds of matter. If you do this you may feel you are
deflected from the main purpose, namely to produce fusion. Sometimes, to take a detour, to be deflected, is the most efficient way.
You may find out how symmetric the implosion has been. You find
out to what extent you have approached your goal of fusion. And
while you are doing it you begin to learn a very great deal about
matter.

I would like to propose an experiment about the behavior of
matter at high density, though I have considerable doubt whether it
is feasible. One important property of matter in highly compressed
state is that you set the electrons free, even at low temperature.
You make an electric conductor out of any material. Now, if you
are careful by shaping your laser pulse, you might be able to produce compression without raising the temperature very much. This
way we may find out whether conductivity indeed occurs at high
densities. One such experiment has been performed quite recently in
Livermore, not with lasers, but with more conventional means of
compression. There is evidence that hydrogen becomes an electric
conductor, as Wigner has predicted some 30 years ago, and the transition point lies somewhere between pressures of 2 and 3 million
atmospheres. I am very confident that with lasers we'll get to
higher pressures. But I am not confident that with lasers we'll
find out where conductivity occurs, because in compressing these
little droplets we are producing a highly conducting dilute plasma
around the droplet, and within this conductor to find a dense
conductor will require more tricky measurements.

Let me conclude with an apology. I wish I could predict that
the energy crisis will be solved by laser fusion. I believe that

in the foreseeable future you can do almost anything with lasers, except solve the present energy crisis.

QUESTIONS AND COMMENTS*

W. J. Buyers, Atomic Energy of Canada: I wonder if you would like to comment on the separation of DF by lasers; what mechanisms do you envisage for extracting deuterium before it recombines to a DF molecule?

E. Teller: Please don't take the proposal I made too seriously. I said this only for fun, only for didactic purposes. There are two ways to separate hydrogen isotopes by lasers in the literature. One is to excite the OH vibration in methanol; if that vibration has one quantum, then it will form hydrogen bromide in the presence of bromine atoms. This seems to work. In fact, the difference between hydrogen and deuterium frequencies are so great that it can even be done with more clumsy methods than with lasers.

The other method is to excite formaldehyde to a UV level which then dissociates into hydrogen and CO. Here the dissociation products are stable enough so that you don't need to catch them. These are the two methods which are truly hopeful.

The one that I mentioned is futurology. You are, of course, quite right that even if you dissociate the molecule you still have to have a hydrogen acceptor. I don't think that will be a difficult question, but the dissociation of hydrogen won't be at all easy.

H. Schwarz, Rensselaer Polytechnic Institute: I have a question in regard to the energy crisis we are in now. You said that it will not be solved by laser fusion and I believe that this is true for the present crisis. Now, how will it be solved?

E. Teller: First of all we'll begin to solve it when the environ- mentalists will recognize that heating our houses might be more important than looking after the love life of the moose in Alaska and, therefore, the Alaskan pipeline can be constructed. This is one step. There are a number of similar steps.

I believe that we ought to speed up the construction of safe nuclear reactors and I don't believe that we need fast breeders. Many other reactors, for instance, the heavy water reactor in Chalk River is a very good candidate because it could run on thorium. That, of course, may solve the problem of generating electricity. But this corresponds only to one-quarter of our energy needs.

────
*
 Moderator: H. Schwarz, Rensselaer Polytechnic Institute.

FUTUROLOGY OF HIGH INTENSITY LASERS   9

Other aspects of the present crisis should be solved by using hydrocarbons and coal. It can be done extensively, for instance, by digging up the coal and then gasifying it. Some of my friends say that it can be done underground by using conventional explosives to loosen up the coal, and then perform the gasification process deep underground where you can control the reactions under high pressure. You have to go more than 500 feet underground, otherwise you cannot produce the right conditions.

G. Baldwin, Rensselaer Polytechnic Institute: During this conference the main emphasis was on producing a nuclear reaction with laser radiation. What are your views on producing nuclear radiation by nuclear reactions with laser beams, in other words, gamma radiation?

E. Teller: Your question is about the gamma ray laser. Well, we'll have that in a short time. I think uranium has an isomer which is only 70 volts above the ground state. I don't think it is hopeless to get an inverted population for that state of uranium. It will only take you into the ultraviolet; it is gamma radiation "honoris causa" because it comes from a nucleus. The high energy gamma rays will be exceedingly hard to get. I think the high energy x-ray laser is much closer. I don't know how it will be done, but it might be done by ionizing atoms and allowing them to re-combine. One may use carbon or aluminum atoms. Coherent x-ray radiation could make a great step forward in analyzing crystals because we could then measure not only intensities but phases and determine the structure of the protein molecules in a routine manner. It will have enormous importance for the whole field of biology.

Incidentally, I talked about pulling apart atoms and molecules. Pulling apart nuclei needs intensities 10^{20} higher. And I don't believe we'll get it.

What about pulling apart the vacuum? Scattering of light by light has a cross-section of approximately 10^{-66} cm^2 and that is too small, even for the highest intensity lasers. But if you have high intensity x-ray lasers scattering x-rays by x-rays will become more feasible, because the cross-section depends on the 6th power of the frequency. A completely new field of physics might be opened.

H. Hora, RPI and Max Planck Institute, Garching, Germany: We talked this morning about the problem of producing electron-positron pairs or proton-antiproton pairs in the laser focus. The advantage of the laser would be that one can draw from laser intensities of 10^{20} to 10^{22} W cm^{-2} higher rates of pair production, 10^{12} higher than with accelerators. My question is: what thoughts do you have on how to store such antiparticles?

E. Teller: I must say that if you want to go into the mass production of anti-matter I wouldn't do it with lasers. I think the
efficiency is likely to be much too small. Using the laser for the
production of pairs, electron-positron pairs or proton-antiproton
pairs will lead to an exceedingly sharp time definition. I do not
know what experiments would benefit by such sharp time definition.
I know that x-rays, sharply defined in time are valuable for taking
precise pictures of the compressions produced by lasers.

   I must admit I have not been able to come up with a high energy
physics application of the laser pair production, although I shouldn't
be surprised if such an application would be found. In this respect
we have too much competition. We can produce these pairs by
accelerating either protons or even whole nuclei to the velocity of
2 BeV per nucleus. It will be really difficult to compete with the
high energy physicists.

Question: What are your thoughts on achieving nuclear fusion energy
with means other than laser fusion?

E. Teller: That I can answer in very few words. The approach of
confining a plasma in a magnetic bottle is promising. Incidentally,
this approach can benefit from laser action which could produce a
dense plasma. I believe that we'll have that in the year 2000 on
a commercial basis, and I hope to have it much sooner as a demonstration. However, the present energy crisis will be solved by reactors
and by better exploitation of coal and of hydrocarbons like oil
shale. The next energy crisis will be solved by conventional fusion.
The energy crisis after that will be solved by laser fusion.

H. Schwarz: Dr. Teller, we all were waiting for the statement in
your last sentence. Thank you very much for your stimulating and
very interesting talk.

**Edward Teller
Monterey, California 1991**

LECTURE IN CONNECTION WITH THE

EDWARD TELLER MEDAL AWARD

Dr. Edward Teller

Lawrence Livermore National Laboratory
P. O. Box 808
Livermore, CA   94550

Let me start by thanking Heinrich Hora for his valuable work on international cooperation and in particular for getting all of us together here. Tonight, we are giving awards to three outstanding men from Russia, the United States and Japan. All this is also a great honor to one man present here who comes from Hungary.

The emphasis in this particular award is on the use of high intensity lasers in producing controlled fusion. I want first to mention in connection with this, Professor Basov, whose work produced high intensity lasers by starting from no laser at all. The very realization of lasers is to a great extent his own work. The only one who may have priority of many years over him is no less than Albert Einstein who came up with the basic idea of lasers one-half century before they became real. It is a pity that we no longer can ask Einstein himself about the all important connection between particles obeying Bose statistics and the instabilities which play such an important role when these particles are produced in the highest numbers.

It was just in connection with such instabilities that the work of my good friend, John Nuckolls, is particularly important. He calculated these instabilities and made predictions about them and also pursued them in connection with very specific experiments at Livermore. Of course, he made me blush with embarrassment by the things he said about me when accepting the medal, but I have to forgive him for the very specific reason that I suspect him of meaning what he said!

Laser Interaction and Related Plasma Phenomena, Vol. 10
Edited by G.H. Miley and H. Hora, Plenum Press, New York, 1992

The recent great accomplishments of Professor Yamanaka have brought the expectation of success on controlled fusion much closer. It is fantastic, but a fact that he managed to compress hydrocarbons to six hundred times their natural density. What next? To expect practical success in controlled fusion in less than ten years may require an excess of optimism but to make predictions of scientific developments for the next century far exceeds my confidence in my ability to see into the future. All of the recipients and many of those present have contributed and will contribute to the realization of controlled fusion.

Now before descending to practical issues, I want to elaborate on the details of my blackboard to which our Director, John Nuckolls referred. Of course, my blackboard contains my thoughts going back a year or so which I was too lazy to erase, but above my blackboard, you will find the conspicuous presence of the picture of a centipede installed there ten years ago by Sandy Guntrum, one of our charming secretaries. I told her that a centipede is the heraldic animal of any big establishment and should be presented with the motto, *Never let your right front foot know what your left hind foot is doing*; nor . . . . She promptly came back and asked me when I expected delivery of the important object. I told her that imitating the Pope's comment to Michelangelo concerning the Sistine Chapel, I will not insist on any time limitations. She delivered the illustration in a prompt and beautiful fashion without the motto, but with little red shoes on each foot of the centipede emphasizing their divergent actions. One of these shoes she explained to me having a hole in the big toe was to remind us of our parent institution, the University of California. I repeat all this because even if the physical feasibility of controlled fusion is solved, the question of economic feasibility will be up to the establishments of whose existence the centipede keeps reminding me.

Having mentioned these important facts, I would now like briefly to mention a point which is both a requirement for our success and may benefit from our success. This is the connection between international cooperation and secrecy. Let me correct myself at once. I should not talk about secrecy but rather about openness.

The first important sign of real changes in the Soviet Union was, indeed, *Glasnost*. The influence of openness in our scientific work and our cooperation and our general behavior can hardly be overestimated. It is now to a considerable extent up to the United States to help establish openness in many ways and, in particular, in connection with the research on controlled fusion. In a specific way, I would like to mention the use of nuclear explosives in connection with experiments which are usually described by the not-sufficiently-descriptive name of tests. By practicing openness, tests could give most valuable scientific results concerning the effects of high energy concentrations which may be difficult to reach even with lasers. In addition to controlled fusion, these high energy densities could lead to the study of states of matter not otherwise accessible to our experience. It would be, for instance, particularly interesting if high compressions could be produced at lower temperatures. It would be certainly interesting to see what the Curie point would do if ferromagnets are several fold compressed. One might even want to speculate what would happen under such conditions to superconductivity.

But returning to the general question of secrecy, I would like to be practical and therefore refrain from a flat statement which I dare not make that secrecy should be abolished. I want to propose, indeed, that secrecy in technical matters involving the collaboration of many people should be limited to a relatively short period, let us say, for instance, the period of one year. We have in the United States tried to keep secrets for a long period of time. I believe the result was that we managed to confuse our own people but did not manage to keep secrets from our competitors.

There are many who argue that secrecy is necessary to prevent the spreading of nuclear weapons. I claim that secrecy is actually of little or no help in delaying proliferation. But openness would be of great help in preventing secret proliferation.

At this meeting, where we are discussing international efforts toward realizing controlled fusion, we have a valid opportunity to speak up for the principle and practice of openness. What I have mentioned and what I will proceed to some little extent to elaborate, I cannot call a solution to the problem. But my words may serve the purpose of stimulating some thought toward practical measures so that we can strengthen international cooperation by the practice of openness.

Of course, the connection between the proliferation of nuclear weapons and secrecy brings us in direct touch with the hard core of our problem. In this regard, we have learned a lot in recent months in connection with the efforts of Saddam Hussein toward producing nuclear weapons. This problem is now under investigation by an international committee whose findings are both unanimous and striking. We now know that Saddam Hussein has spent billions of dollars on the development of nuclear weapons. His approaches were based on valid information. Indeed, secrecy had little to do with delaying his success.

These fortunate delays appear to be due to the technical difficulties which are particularly important in a country where there is a scarcity of people of sufficient know-how in work requiring high technology. A policy of openness would make it obvious whenever massive work on a secret project is undertaken for a substantial period of time. This would give us a better chance to deal with problems of proliferation before they become acute and most dangerous.

I should like to conclude by repeating the great satisfaction I feel for having my name connected with the international effort toward controlled fusion. I am particularly grateful for the opportunity to speak for collaboration and for openness in the company of those who can support constructive action on the specific subject of this conference and also the more general subject of progess and openness of knowledge.

Photo 43

Awarding of the first Edward Teller Medals, Monterey, CA, 12 November 1991

From the right:C. Yamanaka, J.H. Nuckolls, E. Teller, N.G. Basov, H. Hora

Edward Teller Medal Celebration in 1997

The 1997 Medallists from the left: Jürgen Meyer-ter-Vehn, Guillermo Velarde, Edward Teller, George Zimmerman, and Michael Key

Medallists at the LIRPP13 Conference, Monterey 1997 from the left: Nakai, Miley, Meyer-ter-Vehn, Velarde, Teller, Zimmerman, Campbell, Key, Nuckolls, Hora

Photo 45

Twelfth International Conference on
Laser Interaction and Related Plasma Phenomena
Senri Life Science Center, Osaka, Japan
April 24 - 28, 1995

Recipients of the Edward Teller Medal at the 2nd Inertial Fusion Science and Applications Conference, Kyoto, Japan, September 2001
From the left: Meyer-ter-Vehn, Nakai, Rosen, Campbell, Atzeni, Haan, Velarde, Zimmerman, Hora, Key, Shvarts

Edward Teller
Monterey, California 1993

KEYNOTE ADDRESS: THE EDWARD TELLER LECTURE

Dr. Edward Teller
Lawrence Livermore National Laboratory
Livermore, California 94550

The reason that this meeting is so meaningful is because it deals with an area where there is a real opportunity for international cooperation with scientists attacking the very difficult problem of Inertial Confinement Fusion (ICF), which in the end, offers the expectation of practical results. While this meeting on Laser-Plasma Interactions has been going on for a number of years, the first time that such a strong focus on these issues occurred was two years ago. In the meantime, the world has changed, or I should say, the world continues to change. And the "big change" occurred, as we all realize, in the Soviet Union, which no longer exists. In Russia, the people have shown a determination for much meaningful change for democracy, an incredibly imaginative, courageous change in their way of government and life in general. Today, for all Russians, as well as the whole world, the question is, "How do we follow the opportunities opened by this change, in order to create a more stable, a better, a unified world?"

I will not attempt to answer that question, except for saying one thing. I do not believe that solutions formulated in a few sentences can be successfully implemented. I am an extremist, in the sense of supporting gradualism in change. We must go step by step, and the way by which we can achieve our mutual goals is by cooperation.

What we are doing here, is to establish collaboration in our scientific specialty. How this should proceed, I don't know! But at least, I would like to remind you of an American initiative that was taken 47 years ago. (Please don't count the years very carefully! I may be off a few this way or that!) The meaning of that initiative was to unite peoples throughout the world. Harry Truman, who I believe was a great American president, made in my opinion, one mistake. You may call it terrible, you may call it understandable--it was to use the first atomic bomb in an actual application--and not to demonstrate the power of the bomb first, thereby giving Japan a chance to surrender before more people were killed. But having made that mistake, Truman then recognized it was the responsibility of the United States to have a practical involvement in correcting the situation.

3

4   Keynote Address

He turned to a very good man in the State Department, the future Secretary of State Dean Acheson. While Dean Acheson was a clever man, he knew nothing about atomic energy. But he recognized that fact. Therefore, he turned to yet another clever man, David Lilienthal, who likewise knew nothing about atomic energy. Therefore, he contacted Robert Oppenheimer, who knew practically everything about atomic energy! Within a very short time, he prepared what is known as the Acheson-Lilienthal Report.

And in that report, he said, (In my opinion, his thrust in the report was right. We did not always agree on everything, but on that, we did agree.), "Cooperation wins, openness is what we must have. Knowledge is needed at every stage of the atomic energy process: where uranium comes from, how it is processed, how to construct reactors, how to separate isotopes, and most importantly, who works on each specialty, and who is capable of working on it. All technical people should work together, for openness and for peaceful purposes." This Oppenheimer wrote down; Acheson and Lilienthal agreed. Immediately, the President approved it. This was a radical report. President Truman asked a conservative banker, Bernard Baruch, to submit it to the United Nations. He, in turn, suggested that international cooperation must operate under enforceable rules not subject to veto. In the spring of 1946, the proposal was submitted to the United Nations as an initiative for international cooperation. It was, in the end, fully supported by both the right and the left. Unfortunately, Joseph Stalin vetoed the Baruch plan.

You know what happened after that--the Cold War stopped all chances for cooperation. Now, practically half a century later, I am delighted with the prospects for international collaboration. Now, the world really is craving for cooperation.

At the time of the UN agreement, there were no atomic bombs in existence, except a few in the United States. Today, there are 30,000 here and 30,000 there. And that is clearly something unnecessary. I disagree, however, that a nuclear war would be the end of mankind. That is an exaggeration, and one should not exaggerate. Without exaggeration, I can tell you that a nuclear war between the superpowers would have been the end of the United States and the end of the Soviet Union. That war never occurred! In the end, that led to the opportunity for the people of the Soviet Union to show their courage. The United States initiated a change of attitude which has brought us all to the present state of affairs.

Let me now turn to science. One point of particular interest was brought up by my colleague from Livermore, Dr. John Lindl. He gave us a beautiful account of the "tricks" of indirect compression of ICF targets, whereby great changes in density are accomplished. These are incredible densities, accomplished for an equally incredible short time. This raises a point that is not fully explored in the literature: these very high densities give us entrance into a new world of physics. So far, we have not learned too much beyond what we knew earlier about these high densities. This is partly because it is hard to make scientific measurements in the short time in which these densities are accomplished. I can state that this difficulty will be overcome, however, as better diagnostics are rapidly being developed.

Keynote Address   5

There is another problem: the high densities are accomplished, as a general rule, only at very high temperatures, and what happens at high temperatures is, apart from nuclear reactions, not very interesting, because all particles move independently. Thus characteristic collective phenomena, e.g.superconductivity, do not exist in these new high-density states. But we also know, that in achieving this compression, one can shape the time-dependence of the compression in an appropriate way. If you compress slowly, you can stay at lower temperatures. If at the same time you get into really high densities, the most interesting state of matter may occur.

It is in this direction of pure science that I see real progress, and it is here that I see collaboration. At the same time, I also see a radical opportunity--by using nuclear explosives, we can best study this regime. I believe that the existence of the energy concentration connected with nuclear explosions should be used to explore the pure science of high density and of the new phenomena associated with collective effects in this new state of matter.

The second and last major problem we should consider concerns the world energy supply, its effect on mankind, and mankind's response. Professor Nakai and Professor Dautray provided an essential insight into the energy problem we face; namely, that even today, this "problem" is, in a technical sense, open and available for real progress, especially if we are willing to use fission sources in the near-term.

It is not at all surprising that the Japanese and the French should be talking about this issue. It is in Japan and France that nuclear reactors are taken very seriously. Why that is so, I do not fully know. In France, I do know why . . . the French have a really unjustifiable advantage over everyone else: they are logical! And this logic tells them that the nuclear way is the way to go. Are the Japanese logical, Professor Nakai? I'm afraid that we in the United States, and perhaps Great Britain, are a little less logical, and sometimes I regret it!

Now, what should a nuclear power plant look like? Could the worldwide shortage of electricity be solved in the near future by nuclear fission energy, without waiting for the eventual, but long-term, development of inertial fusion power? The nuclear industry has made some foolish mistakes; as long as there are people around, there will be some foolish mistakes. However, there is a way around this problem, a way to eliminate such human errors in the plant. To be short about it, this requires eliminating people from the operating cycle! You separate the reactor from the operators.

The first step in this isolation is to locate the reactor underground. At the same time, you use a design that provides automatic control. This could be done using a design with a gaseous coolant and a reactor with an inherently large negative temperature coefficient. This approach provides automatic control. If the coolant gas returns to the reactor cooler, the reactor power goes up; if you do not need the power, i.e. do not extract the thermal energy from the gas, the gas temperature returning to the reactor goes up. Then, with a strong negative temperature coefficient, the power goes down. This tends to shut down the reactor, thus providing power control on demand. The key to this approach is that a strong negative temperature coefficient must exist, but that can be designed into the reactor core if the importance of this goal is realized.

6   Keynote Address

The next point about this reactor that I envision is: I don't want to change the fuel during the plant lifetime--elimination of refueling would be the second key design goal. Can a reactor run on one fuel-loading for ten years, or for thirty years? Ten years may be possible with existing technology . . . thirty years' lifetime will require the development of new fuel-rod metallurgy, but this should also be possible. The apparatus might be more expensive than present types, but this cost could be offset by other aspects of the plant design. You wouldn't need the complicated apparatus presently used to refuel reactors, and the lengthy shutdown time and exacting safety procedures associated with refueling would be eliminated. This, combined with the load-following capability, means that you don't need operators. If the users want a little more electricity, the coolant temperature will trend downward, and the reactor will respond without manual changes in control-rod position.

The plant that I envision should be feasible if we use all the "tricks" for elimination of operators. Still, another major problem must be faced in this approach. The fission product build-up in a reactor reduces reactivity slowly as operation proceeds, forcing manipulation of control rods to compensate in current designs. However, we can use burnable poisons to counter this decrease. One approach I would call to your attention is the addition, to the conventional uranium/ plutonium fuel, of some thorium, of which there is plenty in the world. Thorium absorbs neutrons, changing it to U-233, which is an excellent fissionable material. And this is just the approach needed to answer the question, "Can we balance operational reactivity changes?" In other words, give the reactor plenty of fuel to begin with, and put in some thorium as a breeding-like component. The thorium would provide an excellent reactivity balance as burn-up proceeds.

Thus far I have mentioned to you quite a few problems that will be hard to solve individually. Taken as a whole, however, interactions may actually help the situation. For example, if there is a large negative temperature coefficient, if the reactor loses some reactivity with time, we just have to run it a little cooler to compensate for the loss. Much can be accomplished just by careful selection of the temperature range within which the reactor operates.

Now, a couple of more questions: what do we do with the reactor, in terms of radioactivity? This leads us to the "big question," what to do with the reactor at the end of its lifetime? With the system I envision, the answer is simple: shut off the coolant, close off the hatches, and forget it! What will happen? The reactor will melt down, and that is just what we want to happen! Previously, people have always considered meltdown as a safety problem leading to fission-product release to the environment. But with the present hypothetical reactor, let the fission products do what they like! I want to put a metal into the fuel. This metal will absorb enough energy to melt and mix with the fuel and fission products upon shutdown. The design should provide a channel in the reactor core chamber, into which this whole melt can flow. Thus, you produce in miniature, a China syndrome!--but, now on purpose! Not that the melted material goes all the way down to China, or even all the way through the underground cavern where the reactor is located. It will be stopped in a volume like a million cubic meters, 100m x 100m x 100m, because that is a region sufficient to contain the heat produced by meltdown. Then you will have captured the fission products in a mixed metallic, very radioactive substance, which could only be handled with very great difficulty. And if someone will try to mine the material, this would become obvious to the world.

Keynote Address   7

I believe that the economics of such a system would be competitive, based on its inherent simplicity--no complicated arrangements for refueling, reprocessing, or terminal radioactive waste storage. You could let a populous city, such as are being created in the Third World, have such a nuclear reactor plant underground nearby, with safety arrangements that the local citizens could understand: "The plant is safe and untouchable." And if a dictator takes over, he will not be able to turn the fissionable products into nuclear weapons; any such attempt would be very complicated and very obvious to the world.

Thus far I have stressed several problems that mankind currently faces and must tackle. However, I believe that in the 21st century, we will continue to face new, equally difficult problems. I believe in progress, but I believe that progress always creates problems. That is where we humans are different from other forms of life. As we are exposed to problems, we respond by changing, but this creates new problems. Without problems, we could not live! But I hope that our future problems will, to a much lesser extent, involve the problems of getting along with other humans. True cooperation among all mankind is developing. I hope we will learn to work for a mutual advantage, without counting in detail, whether you have, or I have, or he has more than anyone else. I think those of you who are really honest about it will see a very complicated and exciting and interesting world evolving. I hope and believe that the future interest will not be in bombs, will not be in war. I believe the answer to our quest for peace is cooperation to solve the problems that mankind faces now and will face in the future. I believe that is what we in this room are doing as scientific and technical people--providing a key example of true cooperation. Indeed I see this as man's saving grace: to deal with the increasing number of very difficult, but lovely problems--lovely in the sense that such problems are the inherent feature of mankind's continued quest for progress.

Thank you very much.

**Edward Teller
Osaka, Japan, 1995**

Keynote address: Dr. Edward Teller

Dr. Edward Teller

Lawrence Livermore National Laboratory

Livermore, California 94550

Ladies and gentlemen, I will try to be very brief. I have to announce a meeting on collision with asteroids that will take place in a little more than three weeks in Livermore, the Planetary Defense Workshop.

There have been meetings of that kind before; one of the earliest and the best in our sister laboratory in Los Alamos, and another very important one in Russia at Chelyabinsk.

The fact is, we all know about the collision of a big object with Jupiter. We know about a big collision 65 million years ago, at the end of the Mesozoic Age and the beginning of the new age, the Cenozoic.

Here is something that does not occur as often as the hurricane. But when it occurs it's worse, so the danger is comparable. And today we know that we can predict it and there are all kinds of methods, including possibly nuclear methods, to prevent it.

We scientists must understand what is at stake, understand what we can do and use this for a very important double purpose. Number one; prevent a possibly very big catastrophe. Number two; use this as a practice to get together, to understand each other, to get things done together. So, we welcome you all to come and stay with us for the conference in Livermore from, I think, the 22nd to 26th of May.

Now having said this, I want to get back to the great thing that we are celebrating here. I want to first remind you again of this remarkable accomplishment of Campbell in developing the area of laser driven fusion. Lasers have the peculiar ability to concentrate energies that start from a conventional source all the way down to volumes small enough to drive nuclear fusion. Campbell is honored for his very important work in the development and use of lasers that have brought us now to the threshold of ignition experiments.

Now, we come to the work of McCrory. McCrory leads the University of Rochester effort in inertial fusion. And he has made important contributions to understanding the instabilities that are a very hard problem, maybe the main problem in inertial fusion.

Did I mention the work in Russia of Kirillov and others doing experiments with pulsed iodine lasers at a surprisingly low cost? I think many important pieces of knowledge will come from this avenue of work.

And finally, I have to thank Miley not only for his many contributions to the field, but for getting us all together and organize the things.

I have no doubt that ignition will work. I don't know yet who, I don't know yet how, I don't know whether it will be inertial confinement or whether it will be a magnetic ignition. But this development has gone steadily. I have talked with some people who say it will take a long time, while others have promised me and I believe them, that ignition will work, not only some time in the distant future, but before I am hundred years old, and I want to be there.

Award Presentation and Keynote Address by Edward Teller
Monterey, California 1997

TELLER AWARD PRESENTATION
AND KEYNOTE ADDRESS

This summary was prepared by H. Hora, UNSW, based on a videotape summary. Some variations for the exact wording may have occurred in transcribing the talks, but the intent is thought to be accurate.

John Nuckolls, (Chair of the Award Ceremony): I wish to introduce Professor Heinrich Hora who has agreed to present some preliminary remarks.

Preliminary Remarks
Heinrich Hora: Thank you very much, John. It's really a special opportunity for me to talk to you here, since I am representing the immediate past history of this conference. It was started nearly thirty years ago under rather complex circumstances. However, prominent highlights can be mentioned. There was John Nuckolls in the 3rd Conference, and Keith Brueckner and a number of great people, E.G., also from Poland, and a Three-Star General, S. Kaliski, later Minister of Science. From Russia, we had from the very first conference on, Gleb Sklizkov, who was specially sent to us by Nobel Laureate, Nikolai Basov. The international color was there, and we were happy that we always survived with our budget at break even, since we had no financial support from outside. A great event indeed was the last, the 12th conference, when it was organized in Japan by Professor Nakai. The conference received very generous support, since outside money came then in, and it was a great conference. Now, we are back in America, as always before. I must make this special point: I resigned as Director in 1991, and the organizational system of this conference has been on the shoulders of George Miley since 1981. The preceding shoulders from the beginning in 1968, were those of the late Helmut Schwarz from the Rensselaer Polytechnic Institute.

We had good international and U.S. recognition, though we always had to worry about classification . These things are now over, fortunately. But nevertheless, the enthusiasm of the people involved, the enthusiasm also from the classified places, and the enthusiasm from worldwide institutions to look into the fascinating laser interaction, was inspiring us and moving us forward. After Helmut Schwarz, we were so grateful that George Miley was ready in 1981, because he had participated in all the conferences from the very beginning, to take over the organization at the American base. He did a marvelous job thanks to his extremely well-organized office. I cannot imagine anything better than what transpired there. This also worked well thanks to his scientific contributions in all directions on the topics of the conference.

Then, from 1991 on, thanks to the committees – it was not a question of one single person – I repeat, thanks to the committees of this institution that it was proposed to Professor Edward Teller, the Great Teller, to combine our organization with the Edward Teller Medal,* and we are most grateful that Edward Teller agreed with this. From that time on, we have been connected – we are connected for all time – and it is a commitment. The commitment is also to Livermore. However, since I learned that NIF is a combined American effort from all the ICF activities, I feel that all the complexities behind this are now over. I am most grateful that Mike Campbell was ready, enthusiastic and fighting for the good of the future to take over the directorship. I thank you so much. And at this point, George Miley for fully understandable reasons, is going to resign from this position and place it into the hands of Mike Campbell and the committees again with their input of ideas, stimulations, inspirations and especially international components for this task.

I should not forget the support by Robert Dautray, now the Haute Commissaire of the French Atomic Energy Commission (CEA), who was always so importantly involved with this business. There are also many others to mention such as Chiyoe Yamanaka in Japan. Also in thanking, I am so grateful that the conference series could move in this direction and will continue along in the same way. I am also grateful that John Nuckolls and all the great names are taking over and working for this, as I have been in the background for some time.

George Miley who worked so hard to push this forward now deserves to place this task into the hands of the others. I would like to thank him very much and ask all of you to express this in the proper manner to George Miley (applause).

* Based on the constitution resolved by the Advisory Board, the medal is awarded in agreement with the American Nuclear Society which opened an account for establishing an endowment fund. For information contact Dr. W. Hogan at LLNL. The recipients of the Medal are: N. Basov (Russia), E. M. Campbell (USA), R. Dautray (France), H. Hora (Australia), M. Key (England), G. Kirillov (Russia), J. Lindl (USA), R. McCrory (USA), J. Meyer-ter-Vehn (Germany), G. Miley (USA), S. Nakai (Japan), J. Nuckolls (USA), G. Velarde (Spain), C. Yamanaka (Japan), and G. Zimmermann (USA).

Response

George Miley: I feel that the time I spent as director of the conference is a time of transition. It is like the transition from fossil fuel to fusion. If you take a long view of history, the fossil fuel area looks like a little "blip" in mankind's history. I came into this meeting on the footsteps of Heinz Hora and Helmut Schwarz, who had such foresight to initiate and nurture the meeting. In those days at the first meeting, we were discussing topics such as – what are hot electrons? – how are ions accelerated in laser interactions? Now, in this meeting, we are talking about relativistic electrons being performed for doing fast ignitor experiments, and we are talking about acceleration of GeV ions! It has been a rapid transition. I feel so good about the meeting we had today among the Advisory Committee. Recognizing the changes that are taking place, the Board feels that now may be the time to move the meeting more squarely into the camp of ICF and its associated applications. This is the plan being discussed and it makes sense. There is a niche in this area for the future which should ensure that the transition is successful for all of the community and all of you who are active participants. I look forward to contributions to the transition through the advisory committee and its director, Mike Campbell.

Response and Introduction of Awardees
Mike Campbell: First, I was honored to be asked by Professors Miley and Hora to take over the direction of this conference. I think George did a wonderful job over the past years. I have altogether too many meetings and too much work to do, but I agreed to this because I want to continue some traditions that I think are so important that this conference has established. One, it truly is international. If I look out into the audience, there are more people from outside the United States than from within. This is wonderful, because the future in science will be more international, because to do the challenges of the next century requires more resources than any one country possesses either in funds or intellect. This conference has its international flavor and I see many friends from Russia, and from Yugoslavia, whom I have not seen for many years, and I was worried about what might be happening with all the problems that exist there. It is great to see them here and science as a good opportunity. I see friends from Japan, China, all over Europe, and from Canada. This is one of the great flavors of this conference, and I think that George and the committees have to be commended, because none of the conferences I have attended have such an international flavor.

I also think this conference is very good because it reflects excellence. For example, the Teller Medal, that Dr. Teller was so generous to associate his name with. One of the four honoraris today, Mike Key, now an employee in Livermore, is from England. Livermore is a microcosm now, I think the changes have happened now in inertial confinement fusion, because English – at least the American English, not the Queens English – is not a language I hear all night walking along the halls of Livermore. We have a Russian scientist growing the crystals for NIF, who we would not have elsewhere. We have a scientist from the General Physics Institutes working at the fast ignitor. We have Joe Kilkenny and Mike Key working on this project. We have Sergei Rubenchik working on laser damage and everything else, as all good Russian theoreticians are able to work. We have Wer Nican from Belgium who helped ensure that the beamlet of NIF works. This conference honors Teller award winners who have demonstrated this excellence I'm talking about. First, there is Mike Key, who is being recognized as an excellent experimentalist.

George Zimmerman: When I first came to Livermore as an experimentalist, I would have loved to prove the codes that were wrong. I have learned and now I see the fallacy of this argument. The only reason we can carry out a one-billion dollar experiment like NIF is due to the codes that George and his colleagues have developed. The Lasnex code George has developed and pioneered is unique. We have strived to make computation physics an integral and an important part of the progress and George is as much a hero of NIF as he is to the code to make MJ target design.

Professor Velarde is honored for being a pioneer in inertial fusion energy – when it was not popular to talk in the United States – for all the work he has done in making Spain a player in this community.

And last, but not least, we honor Professor Meyer-ter-Vehn. I still remember back in 1980 seeing the Meyer-ter-Vehn simple model explaining all our more detailed computations. The work he is now so interested in is the fast ignition. He can always show the richness of the field whether it is direct or indirect drive – there is not one drive against the other. It is increasing the chance for success as well as the fast ignitor. I think this conference enabled these people to be honored, and it is a wonderful aspect and an honor for me to help in continuing this.

So, at last I would like to say what David Crandall was saying yesterday when talking about NIF. For fusion this is a difficult time, even more difficult than in the early 1960s when everyone tried to do something that did not work until the Russian tokamak gave a lead to the field. However, we have an opportunity. We all should give our blessings and prayers, to whomever we pray, that we achieve what everyone has talked about over the last fifty years: for NIF to see ignition. I cannot wait to see the day of the ignition experiment. It is not a Livermore experiment, it is a joint effort of Los Alamos, Rochester, Sandia, Livermore, and of the international community; and all of this happens with ignition With it, I am handing the microphone over to the next speaker.

John Nuckolls: Thank you Mike. I think we are all tremendously fortunate to have Mike in charge of the largest ICF program in the world. And I am sure with his great leadership that he can guide all of us to the top.

Next, we like to move on to the awards part of this ceremony. We are most privileged and honored to have Edward Teller, himself, here at the ceremony. Edward is celebrating his ninetieth birthday next year. He is going to present the medals tonight.

Edward Teller: Ladies and Gentlemen. Mike Key, I am pleased to present the medal to you. We are all tremendously impressed that you go to higher and higher energies, and God knows what all is waiting for us there. Congratulations.

Jürgen Meyer-ter-Vehn, it is a great pleasure to give this to you. I must confess I have the greatest regret of not understanding everything that you are doing. Of course, you are doing theory, and it is theory I should understand. My impression is that you either have explained everything or nearly everything. Congratulations to you.

Professor Velarde, I am not the first one to say it, but perhaps the second one, or perhaps the one-hundredth one to think it - science must be international. We must recognize what is around us and understand the possibility of putting it to use, which is something that is of benefit to everybody and should be done by the community without hesitation and with practically no secrecy. Sir, you have perhaps done more than anyone in ICF to promote this most important direction. Thank you.

George Zimmerman, you can do anything! You can even make a computer work! Computers are getting to be more and more complicated, and I think that they have become one-thousand times better at least, maybe ten -thousand? The point is that we are still not using them to the maximum. You are using them better than anybody I know. The greatest opportunity lies there. Computers have changed in time and we should make even greater use of them. You fully deserved the medal for your efforts.

John Nuckolls: And now we have the pleasure of the after dinner speech from Edward Teller.

Edward Teller: Well, I will tell you something you may not be fully expecting. I hope that in two years from now, four years from now, and twelve years from now, we can do some of this new physics of high density and lasers. Whether we will do it in a competitive fashion, that is a big question. However, in the meantime, I see a technology developing in connection with lasers. I see a technology that can be understood to concentrate energy into the size of less than a millimeter and very clearly by which you can move things. I see a number of new instruments, powerful and sensitive, and very small. I think this is opening opportunities in engineering such that none of us can see the limits for advancement, and I believe it will become very beautiful!

I go on to mention a second aspect. What happens in any material interacting with light, we know. We have theories that are good, simple and beautiful. Actually, they coincide with a theory that we are doing here with lasers. We verify in some terms what we plan for fusion and some is not public. What great confidence we have in the theory of the conditions of high temperatures and densities to realize - I talk not about densities as in the center of stars - but I talk about even higher densities. But here comes the point where our present theories are not very useful, and where something new will have to be learned. That is what we need to understand matter at high pressure and high density, higher than we have today, maybe one-hundred times higher. In particular, this brings up the area of very high densities and low temperature. This is where the theory has to be developed and where something new can be learned. We will open a brand new field which will give new knowledge.

Most of us know the state of ferromagnetism. I want to know how other materials are becoming ferromagnetic, or not becoming ferromagnetic. We have some particular information about what d- and even f-electrons can do in ferromagnetism and high pressures, and this will become even more exciting in new configurations.

However, ferromagnetism is not the whole of the phenomenon.

High-temperature superconductivity may be even more significant. I want to know how the Curie point changes. We have found more and more peculiar data about what d-electrons do in new configurations.

Please realize one point about which we know very little: the condensed state at low temperature and high pressure. What happens at densities of one-hundred and one-thousand times the usual densities of solid state?

There are also the questions of the predictions of the Thomas-Fermi model and how this will reproduce observations at the aforementioned conditions.

The Thomas-Fermi compressibility is one of the new fields of physics. Matter at high density is to be defined properly to ensure that all of the small dimensions of time and space are available for characterization. I think there is a tremendous field of physics lying ahead of us. However, there are only a small number of people presently in the field who can do the job. We need to bring younger scientists into this remarkable area.

And in view of the great opportunities that lay before you, I predict a great future for you and for this physics.

I will ask you whether you have any questions.

Hora: I am grateful that you mentioned the d-electrons from ferromagnetism, as cobalt, iron, nickel, etc. Teller: You forgot manganese! The magnetic and electron properties of Mn-La-Cu-oxides are surely most interesting and must be due to the behavior of d-electrons.

The discussion went on about the magnetic anapol recently measured as predicted by Victor Flambaum. Teller agreed that this is very interesting.

Campbell: You talked about the influences of the public on science. Nuclear fission has not been developed like you would have thought thirty years ago because of the public perception of the problems of nuclear fission. However, what could we learn from this development of fission, what can fusion learn from this?

Teller: I thank you for this question. I am very much aware of this topic. I want to tell you what is wrong. What is wrong is not fission. What is wrong is the fear of novelty. The people have to agree about the subject. At the moment when fusion succeeds, people will be afraid of it, too. I believe the problem is to better explain to the public what we are doing. It is a job - and an unlimited job - to understand how to do this. In the process we must differentiate between the use and the misuse. Let me mention an impression I think you may share. We now have this strange phenomenon known as cloning. The technique to go from one cell of a living being to a whole living being which is succeeding now. It does not succeed ninety-nine times, but it succeeds sometimes. We should explain the opportunities. Our job is to understand how cloning can be used and how it can be misused, to understand how to apply it and to what extent. How to choose what to do with what we have discovered is not to be decided by the scientists. It is to be decided in a democracy by the people (or rather their representatives) informed by the scientists.

Our job as scientists is to justify and promote knowledge. Our job is to find and explain knowledge and give people the courage to use science in, hopefully, a good way.

Laudations to Awardees 1991-1995

PRIOR EDWARD TELLER AWARDEES
1991-1996

Nikolai G. Basov, 1991

As co-inventor of the laser, honoured by the Nobel Prize 1964, Basov was involved with the concept for using the laser for inertial confinement fusion energy from the very beginning. He presented the first publication on this topic in 1963 and reported the first laser produced fusion neutrons in 1968. As director of the Lebedev Physical Institute of the Academy of Science in Moscow he organized the large-scale research there where big laser systems were leading in research and unique scientific results e.g. on compression, DT neutron detection, and second harmonics generation. He essentially promoted related research in his country and on international level through steering conferences and through the IUAEA and UNESCO.

E. Michael Campbell, 1995

Following his undergraduate studies and graduate education at the University of Princeton his carrier was devoted to inertial confinement fusion and laser-plasma interaction. He developed the nuclear activation tracers, which played a key role in verifying the first high densities (100 times the solid state) in laser driven ICF targets. He carried out the first experiments on hohlraums, demonstration of improved coupling with the frequency doubled and tripled laser emission from the Argus laser which he led to the Nova laser at Livermore National Laboratory. Being appointed Associate Director of LLNL he convinced the community to the decision for the National Ignition Facility (NIF). He pioneered sub-picosecond laser-plasma interaction physics which led to his contribution to the discovery of the "fast ignitor" scheme of inertial confinement fusion energy. He also was co-inventor of the target design trhat demonstrated the first laboratory x-ray laser.

Robert Dautray, 1993

His pioneering contributions on energy generation by inertial confinement fusion are based on his long years involvement in all kinds of nuclear physics including: gas flow in isotope separation plants, design and construction of the first material test reactor in Cadarache/Franc and the Franco-German high-neutron-flux reactor in Grenoble (ILL). His contributions are also based on his success in research on high-density and high-temperature plasmas for thermonuclear fusion, leading him to the top position as "Haute Commisaire" of the CEA, The French Atomic Energy Commission. He pioneered laser fusion research and directed the development of the Phebus laser, the most powerful laser in Europe where most significant results on compression dynamics, x-ray properties etc. were gained.

As member of the French Academy of Sciences he directed a key study of climate change due to the greenhouse effect.

Heinrich Hora, 1991

He is known to the ICF community by his long years editing of "Laser and Particle" at Cambridge University Press and from founding and co-organizing the international conferences "Laser Interaction and Related Plasma Phenomena." His optimized adiabatic (uniform) compression scheme for ICF since 1962 led to a gain formula, which is algebraically identical to Kidder's later ρR-formula. He derived the first ponderomotive force including the dielectric plasma properties and formulated the final transient nonlinear force for ponderomotion. His first ponderomotive and relativistic self-focusing formulas have no sub-relativistic limitation. His numerical discovery of the inverted double layers in plasma led to alternative resonances and second harmonics, and to suppression of pulsation at direct drive by smoothing. His adiabatic compression scheme led to high gain volume ignition for ICF.

Gennady A. Kirillov, 1995

The biggest laser-fusion research center outside the United States was established with his continuous contributions and is under his leadership at the Federal Russian Research Institute ofr Experimental Physics at Arzasmas 16 where he is Deputy Director of the Institute. He has developed various laser systems where the biggest for ICF is the ISKRA-5 following the earlier work of K. Hohla and S.B. Kromer. This photochemical iodine laser system produces pulses of 120TW power and 250 ps duration. Consequently the indirect drive ICF experiments produced 10^{10} neutrons from DD filled capsules reaching 7 keV ion temperature. The very detailed analysis of the laser interaction with plasmas arrived at unique results about the stochastic picosecond pulsation. Other very high power lasers are driven by chemical explosives.

John D. Lindl, 1993

From the time of his Ph.D. at Princeton University, he is continuously involved with laser fusion. Even in Princeton he began with his significant contribution endorsing the dielectric plasma effects in the ponderomotive force of laser-plasma interaction with his first clarifying paper based on numerical examples. From then on he worked at the Lawrence Livermore National Laboratory concentrating on fluid instabilities, and elaborating the conditions for high-gain inertial confinement fusion targets both for driving by lasers as well as driven by ion beams. He pioneered the research on implosion symmetry, hohlraum design, high-energy electron production and plasma evolution in the hohlraum, and the physics of compression and ignition. His many contributions are views as providing the basis for our understanding of high-gain ICF targets based on spark ignition of hohlraum targets. In 1990 he became ICF Target Physics Program Leader.

Robert L. McCrory, 1995

After his Ph.D. and prestigious positions at the MIT, he worked on laser-plasma interaction beginning with his position at the Los Alamos National Laboratory in 1972. He made numerous contributions to inertial confinement fusion beginning with his work on the wavelength dependence of the hydrodynamic efficiency of laser driven targets and the hydrodynamic stability theory. Since taking over his position as Director of the Laboratory for Laser Energetics at the University of Rochester in 1981, he continued leading this prominent center for laser fusion to world renown success. In 1988, cryogenically cooled capsules were compressed to 200 times the liquid density of deuterium tritium. His Omega Upgrade 60 beam laser with 30kJ (351nm) pulses produced 2×10^{14} DT fusion neutrons corresponding to a core gain of 30% for break even.

George H. Miley, 1995

He pioneered both, laser research as well as inertial confinement fusion based on his broad involvement in many other kinds of fusion technology. He was the very first to use high-energy electron beams from a pulsed-power diode to directly pump a laser. From his proposal in 1963 for nuclear-pumped lasers he developed with his students a unique Ne-N_2 nuclear-pumped laser, the forerunner of a series of so-called "impurity" nuclear pumped lasers including the important carbon lasers. His first visible nuclear-pumped laser, using He-Hg, led to very low threshold lasers of which the He-Ne-H_2 visible laser has a record-low threshold. In fusion energy he pioneered the low neutron emission options of which advanced fuel is d-^{3}He or p-^{11}B. He proposed the concept of DT spark ignited advanced fuel laser fusion targets. His search for alternate approaches to fusion led to inertial electrostatic confinement as a unique portable low-level neutron source.

Sadao Nakai, 1993

He is involved in laser- and particle-beam interaction with plasmas with the aim of fusion energy since the late sixties and achieved several highlights in this research. His beam device REIDEN was developed for relativistic electron beams or used for proton beams where the highest brightness (10^{14} W/(cm^2-radian)) was reached. This experiment led to the discovery of an anomalous increase of the stopping length for protons by more than a factor ten. He developed the ultra-compact carbon-dioxide laser LEKKO, which in combination with his high intensity electron beams led to a unique high gain near-visible free electron laser used also for laser acceleration of free electrons. Under his directorship of one of the very big world leading centers on laser fusion at Osaka University from 1978, laser compression of polyethylene to 1000 times the solid state was measured and the laser fusion of DT reached the record of 10^{13} neutrons in 1989.

John H. Nuckolls, 1991

Working with Edward Teller the "father of inertial confinement fusion," John Nuckolls, in the late 1950's pursued the ignition of controlled thermonuclear explosion of the smallest possible size using the radiation implosion scheme developed by Teller. With the discovery of the laser in 1960, he began supercomputer calculations of small radiation implosions driven by high power lasers. Temporally shaped (tailored) multi-megajoule laser pulses were used to implode "bare-drop" DT targets to superhigh densities with ignition at a central hot spot. His discoveries were disclosed in a paper with co-workers (nature 239, 192 (1972)) which marked a turning point initiating large scale experiments on laser fusion. He also was the first to apply these computations to heavy ion beam fusion. As leader of this research – finally as Director of the Lawrence Livermore National Laboratory – he initiated nuclear explosive experiments for demonstrating that relatively small energies ignite small masses of DT fuel.

Chiyoe Yamanaka, 1991

He is the founder and first Director of the Institute of Laser Engineering and the following Institute of Laser Technology at the Osaka University in Japan. Since the end of the 1960's he performed leading experiments on laser interaction with plasmas and on laser fusion. His activities led to the second largest glass laser system in the world up to the beginning of the 1990's where most efficient teams in experiments, technology, theory, and numerical studies produced most significant results. Polyethylene was compressed to a density exceeding 1kg/cm^3. The highest neutron yields with direct drive were achieved in the 1980's discovering the experimental proof of an almost adiabatic compression without stagnation or shock waves. These experiments verified the effectiveness of maintaining ideal adiabatic volume compression. With his laser generation of isomeric^{235}U in 1979, he verified the first laser excitation of nuclei.

LAUDATIONS 1999-2003

FROM CITATIONS AND NEWS RELEASES

FOR THE EDWARD TELLER MEDALLISTS 1999-2003 FROM THE INTERNATIONAL "INERTIAL FUSION SCIENCE AND APPLICATIONS" CONFERENCES IN CONJUNCTION WITH THE AMERICAN NUCLEAR SOCIETY AS SUCCESSOR OF THE SERIES "LASER INTERACTION AND RELATED PLASMA PHENOMENA 1969-1997"

Citations for the 1999 recipients of the Edward Teller Medal

Larry R. Foreman, Los Alamos National Laboratory, has excelled as a leader and scientist in the U.S. program to develop and fabricate extremely high quality cryogenic targets for Inertial Confinement Fusion, including targets for the billion-dollar-scale National Ignition Facility now under construction at Lawrence Livermore National Laboratory. His outstanding work has been recognized with a Los Alamos Distinguished Performance Award and a Department of Energy Recognition of Excellence Award.

Steven W. Haan, Lawrence Livermore National Laboratory, is recognized internationally for his outstanding work in the design of Inertial Confinement Fusion targets. He is best known for targets designed for nuclear- and laser-initiated experiments and for developing a first principles model of weakly non-linear hydrodynamic instabilities. With colleagues, he has shared awards from the Naval Research and Lawrence Livermore Laboratories, and the American Physical Society. Haan is a fellow of the APS, cited for "pioneering work in the theory and modeling of hydrodynamic instabilities and mix in ICF targets, and for leadership in the design and analysis of ignition and gain in ICF targets."

Dov Shvarts, Ben Gurion University and Nuclear Research Center of the Negev, has completely changed the understanding of the results of three-dimensional simulations through his work on hydrodynamic instabilities. Shvarts was recognized for his outstanding contributions in receiving the A. D. Bergmann Award for Outstanding Scientific Achievements from the Israeli Atomic Energy Commission. Shvarts is a fellow of the APS, cited for "penetrating insights in the development of theories for ion and electron transport, high-Z opacity, and multi-mode, non-linear mixing due to Rayleigh Taylor and Richtmeyer-Meshkov instabilities."

News Release American Nuclear Society June 26, 2001
Edward Teller Medal Winners Named

Today the American Nuclear Society (ANS) announced the winners of the society's Edward Teller Medal for 2001. The recipients are Italy's Professor Stefano Atzeni of the University of Rome "La Sapienza" and INFM (the Italian National Institute for the Physics of Matter) and Dr. Mordecai Rosen of the U. S. Department of Energy's (DOE) Lawrence Livermore National Laboratory (LLNL).

The Edward Teller Medal recognizes pioneering research and leadership in inertial fusion sciences and applications. The award consists of a minted sterling silver medal, bearing the likeness of Dr. Teller. The medal is mounted on a wooden and inscribed metal plaque. The ANS has established an endowment fund to pay the expenses of the award and welcomes contributions from any source. As this fund grows a cash award to accompany the Medal will be considered.

The awards will be presented September 12 at the Second International Conference on Inertial Fusion Sciences and Applications (IFSA2001) in Kyoto Japan (http://www.ile.osaka-u.ac.jp/ifsa2001/). The Conference, organized by Osaka University, the University of California and Ecole Polytechnique, will bring together about four hundred scientists and engineers from all parts of the world to compare notes on the latest research in inertial fusion.

Professor Atzeni, who did much of the research for this award while he was at the Frascati laboratories of ENEA (Italian National Agency for New Technologies, Energy and the Environment)*, is being honored because of his leading contributions to understanding and teaching the high energy density physics related to Inertial Confinement Fusion (ICF). His contributions to development of the isobaric gain model and to modeling fluid instabilities and mixing are notable. His publications span a broad range of physics issues including implosion symmetry and instabilities, thermonuclear burn, and fast ignition. Professor Atzeni is an outstanding teacher and has lectured on high energy density physics in several European countries and in Japan.

Dr. Rosen of LLNL** is recognized internationally for major contributions to the development of laboratory soft x-ray lasers, and to the design and analysis of complex high energy density and ICF target physics experiments, elucidating electron and radiation transport, and the properties of hot dense matter. These experiments were carried out on a long line of high power lasers at LLNL and contributed to DOE approval of the National Ignition Facility. The American Physical Society has recognized Dr. Rosen's outstanding work by naming him a Fellow, Centennial Lecturer, Distinguished Lecturer in plasma physics, and recipient of its Excellence in Plasma Physics Award. He has also received the Award of Excellence from the DOE and has been appointed as the first Teller Fellow at LLNL.

News Release American Nuclear Society June 26, 2003
Edward Teller Medal Winners Named

Today the American Nuclear Society (ANS) announced the winners of the society's Edward Teller Medal for 2003. The recipients are Japan's Professor Hideaki Takabe of the Institute of Laser Engineering, Osaka University and Dr. Laurance J. Suter of the U. S. Department of Energy's (DOE) Lawrence Livermore National Laboratory (LLNL).

The Edward Teller Medal recognizes pioneering research and leadership in inertial fusion sciences and applications. The award consists of a minted sterling silver medal, bearing the likeness of Dr. Teller. The medal is mounted on a wooden and inscribed metal plaque. The ANS has established an endowment fund to pay the expenses of the award and welcomes contributions from any source. This year's two winners will each receive a $1000.00 cash award from this endowment.

The awards will be presented September 10, 2003 at the Third International Conference on Inertial Fusion Sciences and Applications (IFSA2003) in Monterey California (http://www.llnl.gov/nif/ifsa03/). The Conference, is sponsored by the University of California and organized by seven U.S. inertial fusion laboratories: Lawrence Livermore National Laboratory, Los Alamos National Laboratory, Sandia National Laboratory, General Atomics, University of Rochester Institute of Laser Energetics, Naval Research Laboratory and Lawrence Berkeley National Laboratory. The conference will bring together about four hundred scientists and engineers from all parts of the world to compare notes on the latest research in inertial fusion.

Hideaki Takabe, Institute of Laser Engineering, Osaka University*, is recognized for his pioneering work on laser-plasma interactions, atomic physics, and hydrodynamic instabilities of laser implosions. He was responsible, over a fifteen-year period, for development of the ILESTA code, which is used in Japan like LASNEX is in the United States to design and analyze laser-plasma experiments, and to design inertial fusion targets. Prof. Takabe is widely known for his seminal work on laser plasma ablative stabilization of the Rayleigh-Taylor instability growth. The "Takabe formula" for this process is in world-wide use by physicists and target designers. Prof. Takabe also extended the knowledge of plasma physics and high-energy density physics to astrophysics through collaborations studying the dynamics of supernova 1987A explosion. These collaborations led to the concept of "Laboratory Astrophysics" in which many aspects of astrophysical computational models can be tested in small-scale laboratory experiments. He is now working as Councilor of JPS (Japan Physical Society) to activate AAPPS (Association of Asia Pacific Physical Societies; www.aapps.org) as the world-third pole following APS and EPS.

Laurance J. Suter, Lawrence Livermore National Laboratory**, is recognized for his seminal work on almost all aspects of laser hohlraum physics. Over the past twenty years, he has become widely known as one of the world's leading experts on laser hohlraum physics, with contributions on many topics, including x-ray conversion and drive in hohlraums, symmetry control, the impact of pulse shaping on capsule implosion, and development of a wide variety of experimental techniques to verify and improve the computational models. Dr. Suter's work demonstrated the understanding and control of laser hohlraum physics necessary to obtain approval to build the National Ignition Facility. His recent work shows how to further improve the efficiency and yield of potential NIF ignition experiments, making them even more valuable in the development of high gain IFE targets. His work on hohlraum x-ray physics has also led to the development of novel high efficiency x-ray sources for a variety of other applications.

The Edward Teller Medal is named in honor of Dr. Edward Teller, distinguished physicist, Director Emeritus of Lawrence Livermore National Laboratory and Senior Research Fellow at the Hoover Institution. Dr. Teller is recognized worldwide as a pioneer in inertial fusion sciences. The award has been granted to twenty scientists from ten countries in previous years. It is now under the auspices of the ANS Fusion Energy Division and will be given biannually at the IFSA conferences.

Nominations for the Edward Teller Medal are widely solicited, and nominees need not be members of the ANS. Twenty-three scientists were nominated for this year's medals. Nominees remain eligible for two ballots. The field of candidates was narrowed to two persons by a selection committee comprised of past Edward Teller Medal winners. These recommended recipients were approved by the ANS Fusion Energy Division.

The American Nuclear Society is a not-for-profit, international scientific and educational organization. It was established at the **National Academy of Sciences** in Washington, D.C. in 1954 by individuals seeking to unify the professional activities within the diverse fields of nuclear science and technology. The ANS has since developed a multifarious membership of approximately 13,000 engineers, scientists, administrators, and educators representing more than 1,600 corporations, educational institutions, and government agencies. The Society is governed by three officers and a board of directors elected by the membership.

*More information can be obtained from Prof. Takabe at takabe@ile.osaka-u.ac.jp.
**More information can be obtained from Bob Hirschfeld, Media Relations, LLNL 925-422-2379, hirschfeld2@llnl.gov or from Dr. Laurance Suter at suter1@llnl.gov.

Editorial note: After the awardees 1999 were formally announced, Larry Foreman could receive the Edward Teller Medal in a private ceremony from the IFSA-Director Mike Campbell only because of his serious sickness and could then not deliver a lecture.

LECTURES PRESENTED BY THE EDWARD TELLER MEDALISTS

JOHN H. NUCKOLLS
1991

EDWARD TELLER MEDAL: ACCEPTANCE REMARKS*

John H. Nuckolls
Director
Lawrence Livermore National Laboratory

Tenth International Workshop on
Laser Interaction and Related Plasma Phenomena
Monterey, California

I am honored to receive this award. It is especially significant because Edward Teller is the father of inertial fusion. Teller's pioneering work in the extreme compressibility of matter, the radiation implosion concept, and the physics of thermonuclear burn are fundamental to the creation of very small scale inertially confined fusion explosions.   Edward also made key contributions by fighting to reduce secrecy and by promoting international collaborations.

In the late 1950's, I began to address the challenge of creating the smallest possible fusion explosions. My first supercomputer calculations of micro implosions/fusion explosions of DT masses as small as one milligram were completed in the spring of 1960, some months before the laser was invented. I believed that very small radiation implosions driven by a beam of energy (e.g., a charged particle beam) projected across an explosion chamber would be the best approach to ignition of small fusion explosions in the laboratory. This "indirect drive" approach minimized the energy beam symmetry requirements and maximally decoupled the implosion from the coupling of the energy beam to the target. In these early calculations, I used a spherical target with a very thin shell to relax the beam power requirement. When the laser was invented in late 1960, we immediately recognized its utility for inertial fusion. LLNL physicists Stirling Colgate, Ray Kidder and I independently calculated various methods of using high power lasers to implode and ignite various fusion target designs. Colgate and I calculated implosions in laser driven hohlraums. Kidder applied a spherically symmetric pulse of laser light to the target without use of a radiation implosion.

---

*Performed under the auspices of the U.S. Department of Energy for the Lawrence Livermore National Laboratory under contract W-7405-ENG-48.

In supercomputer calculations carried out in 1961, I utilized extreme pulse shaping to compress a bare drop of thermonuclear fuel in a hohlraum to super high densities and to achieve thermonuclear ignition and propagation from a small central hot spot. This spherical droplet minimized the target fabrication cost and made possible laser fusion power plants and spaceships (which I proposed at that time).

In the early 1960's, Ray Kidder organized and led the Laboratory's (and the world's) first experimental laser fusion program. This few million dollars/year program continued for ten years—when a greatly expanded laser fusion program was launched at LLNL.

In the 1960's and 1970's, my colleagues and I conducted nuclear tests of similar concepts.

In the early 1970's, my colleagues Lowell Wood, Ron Thiessen, George Zimmerman and I produced advanced computer calculations which suggested that the feasibility of laser fusion might be tested by igniting breakeven fusion targets with multi-kilojoule lasers—much smaller than the megajoule-scale lasers required for practical applications such as power plants. These limiting calculations assumed moderately symmetric illumination symmetry, and near perfect pulse shaping and laser target coupling. Symmetry and fluid stability were enhanced by utilizing an electron conduction alternative to indirect drive. This scheme made direct drive laser fusion targets feasible. These calculations were conducted with the LASNEX code newly developed by George Zimmerman.

In the early seventies I proposed declassification of laser fusion implosions of simple spherical droplets. This proposal initiated the laser fusion declassification process

I proposed exploding pusher targets which made possible experiments with 10 to 100 joule lasers. Variations of these targets yielded predictable and detectable numbers of thermonuclear neutrons in the first successful laser implosion experiments at KMS Fusion and at LLNL.

These developments provided the scientific basis for launching a greatly expanded laser fusion program at LLNL and for initiating the construction of a series of large solid state lasers, including the Shiva laser in the 1970's and culminating in the 1980's with Nova, the world's most powerful laser. This expanded laser fusion program was initiated in the early 1970's by Carl Haussmann and led by John Emmett.

In the seventies, the target design group which I recruited and led, proposed high gain indirect drive targets matched to heavy ion accelerators, developed the first successful laser driver indirect drive targets, and achieved compression of DT to 100 times liquid density.

I believe inertial fusion will continue to play a major role not only in basic physics and weapons research, but also as a future world source of energy capable of raising the standard of living on a global scale, and potentially capable of limiting global greenhouse warming by replacing fossil fuels. Just as the ignition of thermonuclear fusion on a micro-scale revolutionized weapons technology, in the 20th century the ignition of thermonuclear fusion energy on very small scales will revolutionize energy production in the 21st century.

NIKOLAI G. BASOV
1991

**COMMENTS ON THE HISTORY AND PROSPECTS FOR
INERTIAL CONFINEMENT FUSION**

Nikolai G. Basov

Lebedev Physical Institute
of the Academy of Sciences
Moscow, Russia

It is a special favour to be here at a celebration for Edward Teller who was the very first in history to demonstrate a man-made exothermic nuclear fusion reaction. This represented the process of inertial confinement fusion (ICF) on a large scale. Now it is a most important aim for mankind to develop this process into a smaller controllable scale for production of energy.

I am very glad to present today some thoughts about fusion and especially on inertial confinement fusion.

Apparently, one of the most important applications of lasers is for ICF. "Laser fusion fever" started in 1962, and very quickly has become an independent scientific trend in physics and technology of thermonuclear fusion. Now one can speak separately on the history of ICF (see Table I).

The present short review of the ICF history does not involve many of the important theoretical and experimental aspects. Areas omitted for lack of space include: the generation of fast electrons and ions, the discovery of numerous laser-plasma effects, the evolution of the target stability problem, and the competition of long-wave and short-wave lasers, etc. While we will not discuss these issues, they have all been a vital part of ICF devlopment.

Such proportions of a laser light as high energy yield during short time and high flux density make it possible to attain specific energy deposition of 10^{18} W/g. This allows one to reach thermonuclear temperatures in the heated matter simultaneously with ultra high density compression. We paid attention to this fact (together with Prof. O. N. Krokhin) in 1962 (a report at the Executive Board of the Academy of Sciences of the USSR, March 1962), and in 1963 the first theoretical evaluations were reported at the III Conference on Quantum Electronics (Paris). Since that time, powerful laser interactions with matter have been studied extensively both

Laser Interaction and Related Plasma Phenomena, Vol. 10
Edited by G.H. Miley and H. Hora, Plenum Press, New York, 1992

25

Table I. Main Steps in the History of ICF

1963; N. G. Basov, O. N. Krokhin	Proposal to use lasers for controlled fusion
1968; N. G. Basov, P. G. Kryukov, O. N. Krokhin, Yu. V. Senatsky, S. D. Zakharov	Registration of thermonuclear neutrons in laser-produced plasma
1971; N. G. Basov, O. N. Krokhin, G. V. Sklizkov, S. I. Fedotov, A. S. Shikanov	First multi-base laser system "Kalmar" ("Russian Monster") for spherical target compression
1972; Livermore Lab., Los Alamos Lab. (USA)	Starting date for the financing of a National ICF Program in the USA
1974; Lebedev Physics Inst. (FIAN) Inst. of Applied Math.	Concept of low entropy compression of shell targets
1975-1978; FIAN	First experiments on low entropy shell target compression (deuterium densities reached 9 g/cm^3)
1970; Livermore Lab	Launching of 10 kJ Nd-laser "Shiva" (neutron yield reached 3×10^{10})
1978; Los Alamos Lab	Launching of 10 kJ CO_2-laser "Helios"
1979-1980 Inst. of Laser Tech. (Japan)	Concept of X-ray and "cannon-ball" targets
1979; Livermore Lab.	Density of compressed fuel reaches 20 g/cm^3
1981-1982 FIAN	Launching of 108-channel laser "Delfin" (stable compression of high-aspect targets and collapse time 200 km/hr)
1983; Livermore Lab.	Launching of a 20 kJ "Novetta" Nd-laser
1983; Inst. of Laser Tech. (Japan)	Launching of 30 kJ Nd-laser "Gekko"
1985-1989 Livermore Lab.	Launching of 130 kJ Nd-laser "Nova" (fuel density, 30 g/cm^3; neutron yield, 3×10^{13})
1990; Livermore Lab.	Appeal to the U.S. Congress on financing an upgrade laser
1990-1991 Inst. of Laser Tech. (Japan)	~ 600 g/cm^3 matter density reached with "Gekko-12" laser facility

theoretically and experimentally at various international labs including the Lebedev Physics Institute (LPI). Our experimental research resulted in the successful development of both ruby and Nd-glass lasers.

In autumn 1962, we put forward the idea of increasing the ruby-laser power by Q-modulation. Then, research on nanosecond pulse amplification using running-wave amplifiers were started at LPI. This work was aimed at development of laser sources of plasma. Experimental and theoretical investigations of the amplification processes in the saturation regime have allowed us to further reduce the laser pulse duration and to reach lasing powers of some GWs. Studies of the amplification processes in the pulses of a complicated multimode structure has resulted in the formulation of a model of solid-state lasers with passive mode synchronization.

At the same time there have been important developments in the methods for hot plasma diagnostics. These methods include the ability to achieve unique spatial-temporal resolution. Some of those methods, e.g. laser interferometry and Schlieren photography, have now come into routine use.

The observation of first thermonuclear neutrons in Nd-laser-produced plasma was an important early success in our experimental research. A year later that result has been reproduced in Limeil (France), and it proved the possibility of using a laser to drive thermonuclear reactions.

The early 70s can be viewed as the "period of laser target compression." In 1972, a 9-channel laser facility "Kalmar" for spherical target irradiation was launched at LPI. Pioneer experiments with spherical homogeneous targets have been carried out with "Kalmar." We first observed the generation of DD-neutrons and, later, the generation of secondary DT-neutrons. The secondary DT-neutrons provided conclusive evidence of a compressed nucleus. A group of scientists from Livermore, E. Teller's group, put forward an attractive and fruitful physical theory about supercompression using a time-profiled laser pulse at a Conference in Montreal the same year (1972).

In 1974, we proposed (together with the scientists from the Institute of Applied Mathematics) an alternative scheme for low-entropy compression by using a time homogeneous laser pulse and inhomogeneous high-aspect targets. During the next few years a great number of experiments on shell target compression were carried out with the "Kalmar" laser facility, and the results have completely proved our compression concepts.

We understood clearly that, when one uses thin shells, the basic problem arises from the possiblity of compression instabilities. Thus, along with a theoretical-numerical study in the late 70s and early 80s, we carried out experiments on compressing the targets with the aspect ratio ~10^2 by using a 108-channel laser facility "Delfin" (launched in 1982). We reached compression of 3×10^3, proving the possibility of stable compression of such targets.

During the last years, the international effort in ICF

research has gained a new impetus. Thus, financing of ICF in the USA reached 200 million dollars per year. Likewise there is a special ICF National Program in Japan. A European Laser Center is planned to be opened, and one of the goals of this center will be a creation of a laser having the energy of hundreds of kilojoules. We hope that the scientists from our Institute will also collaborate in the international effort in this laser center.

Up to now, we have investigated physical processes, which take place during the target compression and burning. Numerous experimental techniques are used for this purpose, and they allow one to trace the various physical processes involved in the target compression (about 30 techniques have been worked out at LPI, they are different and very complicated). The scope of these modern experimental results confirms the practical feasibility of the next step in ICF: the attainment of 200 g/cm^3 density of compressed gas (for comparison note that the density of solid hydrogen is 0.1 g/cm^3) and target burn ignition at 200-kJ laser energy.

In our opinion, at the present time, the basic problem in the ICF is the further development of the physics of thermonuclear targets, especially in the compression range 10^4. Other issues include, the choice and creation of a suitable efficient driver for the ICF and the engineering-technological elaboration of a thermonuclear reactor design. All of these efforts have the goal of a project for a laser thermonuclear electric power plant which is practically realizable, economically profitable, and safe for people and nature.

CHIYOE YAMANAKA
1991

LASER FUSION RESEARCH IN 30 YEARS:

LECTURE OF EDWARD TELLER AWARDEE

C. Yamanaka

Institute for Laser Technology and
Institute of Laser Engineering
Osaka University
2-6 Yamada-oka Suita Osaka 565
Japan

INTRODUCTION

The laser fusion has made a great progress in the last 30 years. Similarly there has been an active international collaboration to develop the magnetic confinement fusion for power research. Recent inertial fusion experiments on the direct driven fusion at Osaka have successfully got the high fusion neutron yield 10^{13} and the high density compression of 600 times normal fuel density. The electron degeneracy of core plasma is also observed. The recent U. S. Halite/Centurion program informed us of indirect driven fusion news that may give high confidence that high gain inertial fusion will be attainable for less than about 10 MJ of driver. However, the data base is not yet clear to determine the details for high gain. This question can only be solved by a large enough laser facility. The U. S. policy on indirect driven fusion program has discouraged international cooperation in ICF for a long time. However, experimental and theoretical progress in ICF in the international community has suggested that the time has come to eliminate unnecessary restrictions on information relevant to the energy applications of ICF. Now ICF is in the second stage of the development. The ignition and breakeven are in a scope of our program. The international collaboration shall be initiated as soon as possible.

It is the most encouraging event for the international cooperation that the Edward Teller Awards are given on the occasion of the 20th anniversary of the Workshop on Laser Interaction and related plasma phenomena which was initiated by Helmut J. Schwarz and Heinrich Hora in 1969 to the four members Nikolai Basov, John Nuckolls, Chiyoe Yamanaka and Heinrich Hora who have been earnestly involved in the laser fusion research for the last 30 years.

WORLD PROGRESS OF ICF RESEARCH

The benefits from achieving high gain ICF in the laboratory are too important for individual countries to try to solve the physics and technology issues in isolation.

Laser Interaction and Related Plasma Phenomena, Vol. 10
Edited by G.H. Miley and H. Hora, Plenum Press, New York, 1992

5

Since 1974, our continuous efforts have been performed to organize the international ICF community concerning the IAEA activity and other authorities. In 1990, the US Secretary of Energy James D. Watkins addressed at the thirteenth IAEA Conference on Plasma Physics and Controlled Nuclear Fusion Research according to the reports by Koonin's committee of NAS and Stever's FPAC and announced if inertial fusion has promise as an energy source - and I believe that it does - we should pursue that promise with sort of cost-effective international collaboration that marks magnetic fusion efforts such as the International Thermonuclear Experimental Reactor (ITER). This address was welcomed by all the participants. A fundamental change is expected. Now, to show the importance of the international collaboration, the world progress of inertial fusion is briefly reviewed setting particular remarks on the Japanese efforts.

At the Levedev Institute in USSR, subsequent pioneering research on ICF has been performed. It let to the disclosure of laser fusion concepts at the International Quantum Electronics Conference in 1963, followed by a presentation in 1968 of the first detection of fusion neutrons from laser irradiated lithium hydride targets. These progresses led to the development of the first world class research lasers, Kalmar and Delfin, which have enabled the Levedev scientists to carry out forefront research in ICF.

At the Lawrence Livermore National Laboratory the pioneering theoretical and experimental works have been performed in the last three decades. Especially a disclosure in 1972 of the implosion physics at the International Quantum Electronics Conference, Montreal by Edward Teller showed that the laser fusion targets could be ignited with much less energy than predicted there-to-fore if the fuel was compressed up to 100 or 1000 times of the normal density. And also the understanding of hydrodynamic and instability phenomena associated with the strong compression ICF targets were prevailed. Series of the Nd glass lasers, Augus, Shiva and Nova were developed to perform the laser fusion experiments. The major glass suppliers for the large Nd glass lasers are in Japan and Germany. LLNL and also LANL have been performed a lot of interesting works in the ICF research which provided essential guidance for the ICF program in the world. However the US classification policy of inertial fusion especially indirect driven fusion is a crucial problem. It hurts the morale of the US scientists who are unable to take credit for their creative work and often must endure the vexation of seeing nearly identical work published in the open literature by workers in Japan, Europe or the Soviet Union. Classification impedes progress by restricting the flow of information, and does not allow all ICF work to benefit from open scientific scrutiny. According to the patient efforts of international movement of several countries, the world situation for cooperation is changing to promote.

In France, the CEA laboratory at Lemeil built a Nd glass laser, phebus, 20kJ which is now open to the public use. Smaller Nd glass laser facility exist at the University of Rochester, Ecole polytechinique Palaiseau, the Shanghai Institute for Optics and Fine Mechanics and at several other places around the world significant progress has been made over the last five years. Since there is in principle no physics obstacle to achieving sufficient control of illumination uniformity and hydrodymanic instability to achieve high convergence, more intense international efforts should be made for exploring abundant and affordable energy.

ICF PROGRESS AT OSAKA

The scientific works at Osaka is internationally recognized. In particular our contributions to ICF theory and experiment culminated in 1987 with remarkable achievement of record compression densities approaching a kg/cm^3. The Institute of Laser Engineering, Osaka University has been a leader among the world's largest ICF laboratories. This is the statement developed by the Awards Committee for the Edward Teller Medal.

In 1972, the Institute of Laser Engineering, Osaka University was established in accordance with the Edward Teller's special lecture on "New Internal Combustion Engine" at Montreal. And also we had timely the first Japan-US Scientific Seminar at

Kyoto by the Japan Society for Promotion of Science in this year which was an origin of the international collaboration on the inertial fusion research where 30 scientists from the U.S., Germany, Britain, Soviet Union and Japan gathered together. At the meeting, K. Hushimi the President of Science Council of Japan attended and encouraged the participants as saying that some person said the ICF is a dark horse, but contrary to it , the ICF is now at the focus of illumination and of attention, you may even speak about a white horse .... In this way the international race of inertial fusion started. Our research on the laser plasma initiated in 1963 using ruby lasers and Nd glass lasers. The first issue of research was the laser-plasma coupling. The absorption mechanisms were thoroughly investigated a result of which was to propose the anomalous absorption caused by the plasma parametric instability. Nonlinear plasma instability due to laser drive became a worldwide popular subject. We investigated the self phase modulation of laser light by plasmas and also the nuclear excitation by electronic transition due to lasers.

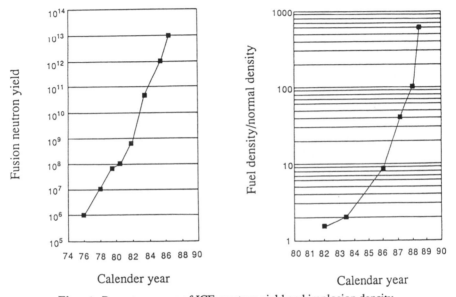

Fig. 1 Recent progress of ICF, neutron yield and implosion density

In 1975, we invented the so called indirect driven fusion concept "Cannonball Target" at our Daisen Summer Seminar which became later the Institute very popular in the world. At the age of oil crisis, the importance of new energy sources was well understood throughout the country. The Prime Minister T. Fukuda who was very favor to fusion sent us a message; "Resources are limited but Human wisdom is unlimited". In a fair wind to fusion research we set LEKKO CO_2 lasers to the Los Alamos group and competed with Livermore people by GEKKO glass lasers and compared the ideas to the Sandia team by REDEN beam machines. In the laboratory different kinds of these three drivers enabled us to compare the results of various plasma experiments and to review the qualification of drivers.

In 1983 the world largest of glass laser GEKKO XII was completed by the cooperation of the NEC and we got a fine estimation of the world leading laser fusion laboratory. As for the direct driven ICF, it is potentially more efficient but has significantly more stringent requirements on driver beam uniformity and the control of hydrodynamic instabilities. We had significant progress in this field using a novel type of uniform shell target and a random phasing smooth laser beam.

In 1985, the new idea of LHART (Large High Aspect Ratio Target) was devised by using an implosion simulation code. It could record a super shot of DT fusion neutron yield 10^{12} which was hurriedly after traced by the LLNL group. In 1986 the Fusion Power Associates gave a Leadership Award to this laser fusion achievement.

In 1987, The green light random phasing 12 beam of GEKKO XII glass laser irradiated a plastic shell target of nearly perfect sphericity to attain the 600 times normal density. The D-T fuel density reached $120 gr/cm^3$ in absolute. The plasma is some what Fermi degenerated. These details were reported at the IAEA conference in Nice at 1988. Ablative pressure generation and hydrodynamic behavior of compressed fuel were experimentally and theoretically investigated. The implosion performance was optimized by using an appropriate aspect ratio of the target. The uniformity of laser irradiation as well as the pellet structure were essentially important to avoid the growth of instability.

Since the Edward Teller's lecture at Montreal, it has passed 20 years to attain the high compression densities of fuel predicted. These results give us high confidence that the ignition and burn of ICF will be attainable with a 100 kJ laser driver, such as GEKKO XII laser up grade.

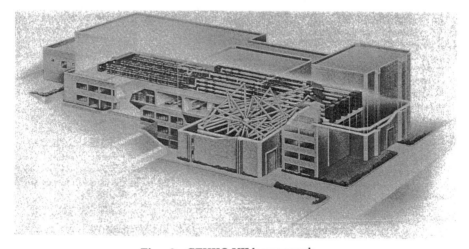

Fig. 2 GEKKO XII laser upgrade

PROSPECT OF ICF

At the Edward Teller Award ceremony, Teller predicted again the prospect of ICF research that the ignition and breakeven will be attained in the next 10 years before the 21st century following the recent results of high compression. He is content with the fulfillment of his first prediction of high density compression by Osaka. We shall say the Prospect of ICF seems to be very bright.

Applications of inertial confinement fusion include not only civil energy production but also physics at the laser-atomic frontier, nuclear matter under extreme conditions, cosmology, special isotope separation, food preservation, hydrogen production and advanced space propulsion. The pursuit of ICF will contribute substantially to overall scientific strength in several areas.

In the international collaboration, the essential research and technology development for fusion engineering and technology should be carried out in the following items,

(1) High-average-power fusion drivers, lasers as well as heavy ion beams. They are also to produce the significant new applications in industry and science.
(2) Power-plant use ICF target and fueling technology including cryogenic methods.
(3) Confinement-chamber materials and energy conversion technology including studies of neutron damage and tritium handling.

The ICF reactor driver development is essential and adding to it, material research including target, tritium and structural materials and energy conversion systems need an intense technology development effort for fusion power plant to be a reality by 2025. The total integrated cost will be about $20 billion to get the first prototype ICF demonstrate power plant by 2025. Thus, after the year of 2000, when the breakeven might be attained, works on fusion energy technology would become heavy and strong. The international center for integrating a demonstration power plant of ICF should be contemplated. No other alternate energy source holds the bright promise of fusion and none has ever presented such formidable scientific and engineering challenges.

HEINRICH HORA
1991

NEW BASIC PHYSICS DERIVED FROM LASER PLASMA INTERACTION*

Heinrich Hora

CERN, CH1211 Geneva 23, Switzerland**

INTRODUCTION

When Einstein's discovery of the stimulated emission was realised in the laser in 1960, a new chapter in physics was opened up. The ability to concentrate electromagnetic energy spatially into the range of the wave lengths, and temporally to the oscillation time of light (including x-rays), has permitted the concentration of light to an intensity 24 orders of magnitude higher than that of sunlight falling on earth and to energy densities of MeV/atom in solid state densities.

It was fully expected that the physics of the interaction of such radiation with matter would reveal completely unpredictable phenomena. The following examples should be considered only as an overture to how physics needs to be generalized and explored further in order to discover the many new phenomena which can be expected in the future.

These examples relate to high intensity interactions of laser radiation with condensed or gaseous materials where the radiation converts the atoms into electrons and ions of a high temperature plasma within a very short time. The interaction conditions are orders of magnitude different from those of the preceeding plasma physics and it is no surprise that a new area of plasma physics had to be developed and explored. All the phenomena considered here are highly nonlinear. Although these types of phenomena were well known in theory, continued experience of them has revealed as a general principle of nonlinear physics, that much higher accuracy is needed in specifying the initial presumptions of any model or theory than was necessary in the case of the earlier linear physics. One consequence of this is that it is now much more difficult to correctly predict results in theoretical physics, however, it is likely that perserverence will lead to the prediction of phenomena which could not now be imagined.

---

* Edward Teller Lecture delivered at the occasion of the award of the EDWARD TELLER MEDAL FOR ACHIEVEMENTS IN FUSION ENERGY, Monterey Cf.12 November 1991
** also from Department Theoretical Physics, University of NSW, Kensington 2033, Australia and Department Applied Physics, University of Central Queensland, Rockhampton 4702, Australia

VARIOUS PHENOMENA

Studies of laser irradiation of gaseous or solid targets initially revealed classical behaviour for heating, ionization and gas dynamics with temperatures of up to 100,000 degrees (10 eV temperature). For slightly higher laser powers (above MW), however, the plasmas generated suddenly revealed ions with energies 1000 times higher than gas dynamic ones. The optical constant (i.e. dielectric response, including absorption) corresponding to conditions, had to be generalized from the classical values to intensity dependent nonlinear formulations which could include relativistic effects. The theory of self focusing of laser beams by ponderomotive forces and the relativistic self focusing was one consequence of such generalization.

The most ambitious application of a laser-plasma interaction is the generation of clean, low cost and inexhaustable nuclear fusion energy. The aim here is to use laser irradiation to both heat and compress a pellet or capsule of high density deuterium and tritium in order to ignite exothermal fusion reactions such that the fusion energy gained is much higher than the laser energy applied. There were two main problems to be solved (a) the interaction of the laser radiation appeared to be extremely complex, with unexpected and rarely understood instabilities, pulsations and anomalies, and (b) the achievement of sufficiently high gains required the ignition of a small core of the pellet (e.g. by a central spark) to initiate the reaction of the surrounding material by a fusion detonation wave.

Many laboratories contributed to providing solutions, but it was due to the triumphant achievements of the Lawrence Livermore National Laboratory in general, and of John Nuckolls [1] in particular that a solution based on an understanding of the physics can now be offered for the establishment of an economically competitive inertial fusion energy reactor at the beginning of the next century. Nuckolls solved the complexity of the interaction by using indirect drive and also by introducing the above mentioned spark ignition to produce sufficiently high gains.

The task of providing a physics solution for initiating a large scale developmental program for laser fusion reactors is now complete. It is now in the hands of the politicians to decide how it should be developed.

In support for this physics solution the following alternative scheme can also be considered [see Chapter 13 of Ref.2] as providing a safe solution which has been experimentally confirmed by underground nuclear explosions during the Centurion-Halite project. The alternative scheme solved the complexity of the interaction by clarifying the main problem of the interaction process (i.e. the pulsation in the 10 to 30 psec range), via a numerical modelling of this process and by the application of smoothing techniques which can suppress the pulsation by the use of short time coherence and the superposition of fields. These methods initially aimed at a smoothing of the laser irradiation in the lateral direction to achieve uniform intensity profiles and to avoid self-focusing and other instabilities, but owing to the experimental studies of pulsation by Maddever and Luther-Davies in Canberra, and our numerical understanding the concept of direct drive with smooth interaction and low reflectivity can be considered.

The achievement of sufficiently high gains (the second problem) is possible without the rather complex and very sensitive parameters required when using spark ignition. The earlier very inefficient volume compression and burn, led us to the numerical observation of a volume ignition mechanism such as in a Diesel engine (1978) with a decrease of the optimum ignition temperature to 4 keV by self heating of the reaction products and to temperatures even of only 1.5 keV, due to self-absorption of bremsstrahlung. The full equivalence of the results of this volume ignition with that of the spark ignition has been demonstrated and some interlinking of the processes have been possible.

According to this view, the fully established fusion reactor concept of Livermore can be supported by the possibility that the otherwise unclear properties of "mix" problems and sensitivity of parameters can be avoided completely in favour of the very simple and safe volume ignition process. The earlier question was whether laser ablation of pellets could produce this ideally adiabatic compression free of stagnation and shock waves. This problem was solved experimentally in 1985 by C Yamanaka, S Nakai, T Yamanaka and co-workers [3] where high fusion gains were measured.

COMPLETION OF THE EQUATION OF MOTION BY NONLINEAR FORCES

One central problem in laser plasma interactions was to find the correct equation of motion for the plasma. The problems were caused from the beginning by the very strong inhomogenities in density and temperature in laser irradiated plasmas. Most earlier theories of plasmas were based on the homogeneous collisionless plasma. Inclusion of collisions was one of the crucial problems at laser interaction. We have shown how dispersion functions from a pole changed from minus infinity for neglection of the collisions into values of nearly plus infinity when collisons were included. The poles of the optical constants near the critical densities especially caused very high gradients for the quantities in the near neighbourhood of these poles. The neglect of collisions led to fundamentally different results in resonances and other phenomena. A quantum modification of the collision frequency at high electron temperatures appeared to be necessary, although all the consequences of this fact have by no means been exhausted.

Prior to 1966, the problems of inhomogenities and collisions in the theory of the equation of motion, were not so dramatic. Spitzer's derivation for the equation of motion from the Boltzmann equation arrived at a force density in a plasma with an ion mass m_i and ion density n_i and a net velocity $\underline{v}$ of the space charge free and thermally equilibrated plasma of:

(1) $$f = m_i n_i (\frac{\partial}{\partial t} v + v . \nabla v) = -\nabla p + \frac{1}{c} j \times H + F$$

which is determined by the thermokinetic force f_{th}, given by the negative gradient of the gasdynamic pressure P, and by the Lorentz force due to electric currents in the plasma and the magnetic fields $\underline{B}$, and additional forces $\underline{F}$ like gravitation etc..

If the currents and magnetic fields are due to the high frequency field (e.g. of a laser), the Lorentz term appeared to be the pondermotive force

(2) $$\frac{1}{c} j \times H = \frac{1}{4\pi} \nabla . E^2$$

This force was derived first in 1846 as electroscriction by Kelvin and later by Helmholtz as a ponderomotive force where the elastomechanical analogues were used to understand the electric field $\underline{E}$. It was to the credit of Erich Weibel (1957) that he could show that electrons in high frequency fields of microwaves obey the same field gradient forces (2) as evaluated by Kibble (1966) from Lagrangeans.

When looking at the inhomogeneous plasmas the gradient of the density is essential, as expressed by the gradient of the optical refractive index ñ. The forces (2) in a plasma with one dimensional geometry were then (Hora, Pfirsch and Schluter 1967 [4])

(3) $$f_{nl} = - i_x \frac{E_v^2}{16\pi} \frac{\omega_p^2}{\omega^2} \frac{\partial}{\partial x} \frac{1}{ñ(x)}$$

given as the time averaged nonlinear force in the plasma due to the propagation of light of frequency ω with an amplitude E_v of the electric field and using the plasma frequency ω_p. This force is identical with the ponderomotive forces (2) as seen from the WKB approximation and stands for the Lorentz terms in (1), expressing now the dielectric explosion (Fig. 1) of a plasma density profile resulting from laser irradiation. This predicted dielectric explosion was discovered numerically as the generation of a caviton by Shearer, Kidder and Zink 1969 [5], based on the nonlinear force formula (3) driving thick blocks of plasma to the observed keV ion energies, and producing the charactertistic density minima (cavitons) (see Fig.2).

Difficulties appeared when the one dimensional geometry was extended to more dimensions (e.g. in the case of laser radiation obliquely incident on a plasma). In

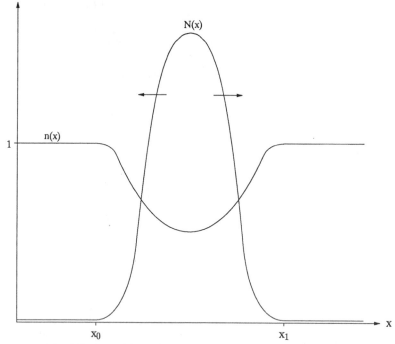

Fig. 1 Plasma with an electron density N(x) and subsequent real
part of refractive index n(x) as irradiated by a laser from the
l.h.s.. Arrows show the nonlinear forces by the negative gradient
of 1/n(x), Eq. (3) [4].

this case the equation of motion had to be generalized such that the nonlinear force
became [6]

$$(4) \quad f_{nl} = \frac{1}{c}j \times H + \frac{1}{4\pi}E\nabla.\ E + \frac{1}{4\pi}EE\ .\ \nabla(\tilde{n}^2 - 1) + \frac{\tilde{n}^2 - 1}{4\pi}E\nabla\ .\ E + \frac{\tilde{n}^2 - 1}{4\pi}E\ .\ \nabla E$$

where all vectors on the r.h.s. correspond to the oscillation of the high frequency field

of the laser and $\tilde{n}$ is the complex refractive index, including the intensity dependent
nonlinear modifications of the dielectric response and of the absorption.   When

differentiating the last term, three terms appear and it should be noted that Schlüter's
derivation of the equation of motion in 1950 reproduced one of the three nonlinear
terms which did not appear in Spitzer's derivation from the kinetic theory. We showed
from momentum conservation that we must use all the terms in our solution (4) to
describe the force density based on non-transient collisionless plasma properties. When
including collisions, the non-pondermotive terms appeared, determining the ordinary
radiation pressure by absorption (e.g. perpendicularly to the density gradient of the
plasma).

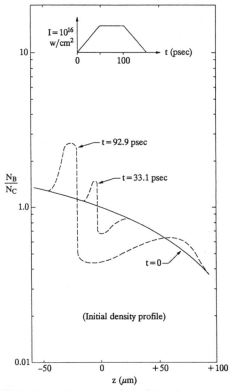

Fig. 2 Hydrodynamic computation of the electron density N_e per critical electron density $N_c = 10^{21}$ cm^{-3} at different times at dependence of the intensity I up to 10^{16} W/cm^3 is given by the upper insert. The monotoneous initial density changes at later times t to profiles with minima (discovery of the caviton) following Shearer, Kidder and Zink [5].

While our Eq.(4) was the final formulation for the time-independent solution, for the transient case several years of controversy occurred before it was decided which terms are needed. Beyond the former approximate solutions, we finally found the following solution (1985) [7]

$$(5) \qquad f_{nl} = \frac{1}{c}j \times H + \frac{1}{4\pi}E\nabla \cdot E + \frac{1}{4\pi}(1 - \frac{1}{\omega}\frac{\partial}{\partial t})\nabla \cdot EE(\tilde{n}^2 - 1)$$

which is algebraically identical with the formulation using the Maxwellian stress tensor for the vacuum T,

$$(5a) \qquad f_{nl} = \nabla \cdot \left[T + (1 - \frac{1}{\omega}\frac{\partial}{\partial t})\frac{\tilde{n}^2 - 1}{4\pi}EE \right] + \frac{1}{4\pi c}\frac{\partial}{\partial t}E \times H$$

15

The correctness and final generality of this nonapproximate equation of motion was first derived by the algebraic structure of the terms (1985) and was later confirmed by T Rowlands from Lorentz and gauge invariance.

NONLINEAR PRINCIPLE

On one hand, it was a basic achievement to derive the complete and general formulation of the force density in plasma theory, on the other hand one had to understand how in laser plasma interaction, the dielectric gradients according to our formula (3) produce the high plasma velocities by dielectric explosion. As seen from the WKB approximation, the electric laser field amplitude $E = E_v/|\tilde{n}|^{\frac{1}{2}}$ in a nearly collisionless plasma increases to very high values. For example, if the absolute value of the refractive index $\tilde{n}$ takes values of 1/10 or 1/100 (or much less in the plasma), the dielectric explosion is then due to the negative gradients of the increased E^2 values, and therefore the ordinary radiation pressure is increased dielectrically by a factor 10, or 100 (or more) respectively.

This could be seen immediately numerically and in experiments by several groups due to the generation of the density minimum near the critical plasma density at very high intensity laser interaction with plasmas. The action of the nonlinear force was checked also in the Boreham experiment [8] where a laser beam was focused into a low density gas and the generated electrons were emitted radially from the beam with energies in the range of 100 eV to 1 keV, as given by half of the maximum quiver energy of the electrons which the nonlinear force converts into translational energy.

While this conversion could be understood immediately from the nonlinear force by global considerations, the analysis of the single particle motion in the laser fields produced discrepancies. The reason was very simple: in the analysis, as usual, the transversal electric and magnetic field of the laser was used. These fields, however, do not exactly fulfill the Maxwellian equations if one considers a laser beam of finite diameter. When we derived the missing field components - it was the discovery of the first exact longitudinal (sic!) components of light in vacuum - their inclusion in the force produced agreement with the measurements.

This taught us that only with the Maxwellian exact solutions could we obtain the correct description, and that neglecting the very small longitudinal components led to a completely wrong result (i.e. a change from yes into no).

This was an example - and others were developed later - of how nonlinear physics cannot be done simply by using a next higher approximation of a second order extension, but one has to use the fully exact linear model or theory in order to arrive at correct predictions.

The old fashioned method of using theoretical physics to predict phenomena is certainly possible in nonlinear physics, but it is much more difficult to keep the correct linear ingredients and to avoid approximations. In this way, however, phenomena can be derived or predicted which cannot be thought of at all in physics.

For example, ion beam fusion is absolutely impossible according to Spitzer (1951) and magnetic confinement was considered as the way to progress. Spitzer was completely correct in his mathematics, physics and logic, but his result was nevertheless wrong: it was linear only. Nonlinear physics, however, does provide a solution for ion beam fusion, at least theoretically. Therefore, realizing the principle of accuracy in nonlinear physics will open completely new dimensions of physical research for the future as a recipe for elaboration and for solving very difficult mathematical problems.

The Boreham experiment [9] was an example for the application of the correspondence principle of electromagnetic interaction [10]. Contrary to Bohr's correspondence principle for the electronic states in atoms of very high fundamental quantum number, there is an easy way for an electromagnetic interaction to continously change from the quantum interaction to the classical interaction [10] just by continously varying the laser intensity for the interaction. While the lower intensity results in quantum interaction as seen, for example, from multiphoton ionization [9], the higher intensities result in point-mechanical behaviour [8], as seen from the appearance of the Keldysh quasi-classical tunnel ionization [10].

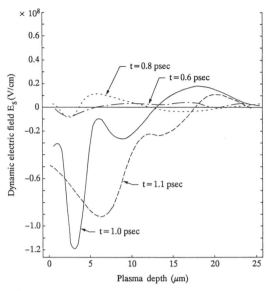

DOUBLE LAYERS AND SURFACE TENSION

Fig. 3 Amplitude $\underline{E}_s$ of longitudinal oscillations of the electric field driven by 10^{16} W/cm^2 Ng glass laser pulse incident from l.h.s. on a plasma slab generating an inverted double layer at the caviton near 5 micrometer [12]

Surface tension of liquids or solids is mostly related to dipoles of the molecules which are not saturated at the surface. Since high temperature, full ionized plasmas consist of electrons and nuclei only, no dipoles of the kind mentioned could be expected, and it would be strange to ask about surface tension of plasmas. The way in which this, nevertheless, is possible, became obvious from laser plasma interactions for studying nonlinear force.

Contrary to the space charge quasineutral theory of plasmas (which is correct only for homogeneous plasma), on which assumption the old two-fluid theory of the nonlinear force of section 3 is based, we knew from a semi-microscopic derivation of the nonlinear force that the light is pushing or pulling the whole column of the electron gas within the space charge neutral ion background, and the ions follow by electrostatic attraction, giving them the inertia for plasma motion. It was therefore necessary to use a genuine two-fluid model in contrast to the earlier two-fluid theory of

Schlüter and Spitzer. This genuine two-fluid model uses separate electron and ion fluids coupled only by Maxwell's equations. This reduces to the Poisson equation only in one dimensional geometry, but generally it results in the spontaneous magnetic fields of Megagauss values in laser produced plasmas in a three dimensional treatment of the genuine two-fluid model.

The electric fields $\underline{E}_s$ from the Poisson equation in plasmas are well known as ambipolar electric fields given by the gradient of any pressure. The electromagnetic field driven ponderomotion may also produce a pressure which causes electric fields in the same way, but generally the nonlinear force is <u>not conservative</u> and (so called "ponderomotive") potentials appear only in very special simplified conditions. Our genuine two-fluid model revealed the general electrostatic oscillations including collisional damping in any inhomogeneous plasma which contains the ambipolar field, but which is of very complex generality, including then the nonlinear force effects and oscillation given by the locally varying Langmuir (plasma) frequency and collisional damping.

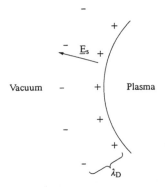

Fig. 4 Plasma Expanding into vacuum. The faster electrons move ahead of the slower (equithermal) ions establishing a double layer of the thickness of the Debye length D with an internal electric field $\underline{E}_s$

With laser irradiation of plasma, a new type of resonance was derived for perpendicular incidence of laser radiation. This was sought for a long time because resonance absorption (Forsterling-Denisov) works at oblique incidence only and is polarization dependent. We further derived strong second harmonic emission in very low density plasma with laser irradiation in agreement with observations.

This all resulted in a very general understanding of internal electric fields and double layers in laser produced plasmas and even the hydrodynamic derivation of the laser driven, very large amplitude, Langmuir oscillation with the Langmuir pseudo-waves was possible. The double layer results were in agreement with significant measurements at the SOREQ Nuclear Research Centre in Israel.

One could immediately see numerically, and as confirmed experimentally, how the cavitons produced by the nonlinear force resulted [11] in inverted double layers (Fig.3). The simple double layer at the surface of a laser-produced plasma expanding against vacuum is seen in Fig. 4. The electrons due to their smaller mass leave the surface where the heavier ions remain. The charge separation causes an electric field and the potential corresponds to the plasma temperature T or - at very intense laser irradiation - to the much higher potential produced by the nonlinear force.

The thickness of the double layer is of the order λ_D of the Debye length. If one integrates the electrostatic energy within the surface area one arrives at surface tension given by this energy per surface area as we derived together with Shalom Eliezer [12]

$$(6) \qquad \alpha = \frac{4\pi R^2 \int (E_s^2/8\pi)dR}{4\pi R^2} = \frac{(gkT)^2}{e^2 8\pi} = 4.75 \times 10^{-16} \, g^2 n^{\frac{1}{2}} T^{3/2} \, \frac{erg}{cm^2}; \; [T] = {}^\circ K$$

where R is the radius of the sphere in Fig. 4, $\underline{E}_s$ is the "electrostatic" field in the plasma surface, k is the Boltzmann constant, and e is the electron charge.

This surface tension can have values of Joules/cm² in laser produced plasmas. It acts against the Rayleigh-Taylor instability in a way similar to the non-disintegrating water droplets. This causes the smooth surface of laser produced plasma plumes. Surface waves of a length shorter than about 100 times the Debye length are stabilized.

This can all also be applied to the degenerate electron gas in a metal. These electrons, similar to those of the plasma, tend to leave the ion lattice in this case not by thermal energy but with their Fermi energy of some eV. They are then stopped by the electric field that they generate as a surface double layer, resulting in the potential given as a work function of some eV, and a surface tension results in the same way as in Eq. (6). In cooperation with R S Pease [p176 of Ref.2], we derived values of the surface tension of metals that give good agreement with the experimental values. These surface tensions were all positive according to the model of the generation of the swimming electron layer above the lattice ions, in contrast to the jellium model of surface tension of metals which can give negative surface tensions contrary to measurement.

CONTAINMENT FORCE OF HADRONS IN NUCLEI AND PHASE TRANSITION INTO QUARK GLUON PLASMA

The surface tension for metals as given by this plasma model (expression (6)) for a degenerate plasma, can be extended to the case where a plasma is no longer defined for compensating charges, namely for a nucleus [13], just by substituting for the temperature T with the Fermi energy. Charges are present and Hofstadter's experiments showed how the charge distribution in a nucleus is constant in the interior, and how it decays over a quite long distance of about 3.5 fm from the constant value to zero at the surface of the nucleus.

We can now use the Fermi energy simply to define a similar "plasma like" surface tension and surface energy for the nucleus in conjunction with the hadron mass (that of protons or neutrons) and the Compton wave length λ_C for the nucleons

$$(7) \qquad E_F = \frac{(3/\pi)^{2/3}}{4} \frac{h^2 n^{2/3}}{2m} \frac{1}{(\lambda_c/2) \, [n + 1/ \, (\lambda_c/2)^3]^{1/3}}$$

$$
E_F =
\begin{cases}
\dfrac{(3/\pi)^{2/3}}{4} \dfrac{h^2 n^{2/3}}{2m} & \text{(subrelativistic)} \qquad (7a)\\[2em]
(3/\pi)^{2/3} \dfrac{hcn^{1/3}}{4} & \text{(relativistic)} \qquad (7b)
\end{cases}
$$

which changes from the subrelativistic branch into the relativistic one at a nucleon density of $(\lambda_C/2)^{-3}$. We note that the relativistic Fermi energy is not dependent on the particle mass and therefore is the same whether the mass is that of nucleons or quarks etc..

The internal energy E_i of the bismuth nucleus is mainly determined by the Fermi energy and to an extent by Coulomb repulsion, a dipole surface energy and a volume energy, to arrive at 4.17 GeV. The surface energy E_s of the bismuth nucleus using the Fermi energy as mentioned in an Eq. (6) produces a value of 4.14 GeV, confirming that our plasma surface tension model leads to a stable nucleus at a nucleon density as measured. We also derived a Debye length for the Hofstadter decay of the charge density of 3.64 fm in good agreement with the measurements.

What is interesting is that the ratio

$$(8) \qquad E_s/E_i = \text{const } n^{1/6}$$

becomes one at the nucleon density n of the stable nucleus. We see that for lower hadron densities there is indeed a surface tension and surface energy, but this does not compensate for the internal energy of the nucleus (Fig.5) [13]. We find further that our surface energy just corresponds to the tangling bond energy of the nucleons given by the Yukawa potential at the surface.

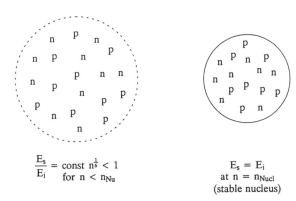

$$\frac{E_s}{E_i} = const \ n^{\frac{1}{6}} < 1$$
$$for \ n < n_{Nu}$$

$$E_s = E_i$$
$$at \ n = n_{Nucl}$$
$$(stable \ nucleus)$$

Fig. 5 Hadrons (protons p and neutrons n) at less than nuclear density n_{Nu} where the surface energy E_s is less than the internal energy E_i. Equality is reached at the known density of the nuclei explaining then how the surface energy is exactly compensating the internal energy for producing a stable nucleus.

What is even more interesting is what happens if the nucleus is compressed as in stars or for example, by the heavy nuclear collision where shock waves produce up to six times the nuclear density (Scheid, Muller and Greiner, 1970). The increase of the nuclear density will then just surpass the relativistic threshold in Eq. (7) and will reach a ratio given by Eq. (8) which does not depend on the nucleon density. The formation of nuclei makes then no sense at all. Since - as mentioned - there is not a mass defined for the particles, these can well be a quark gluon plasma as expected in the interior of dense stars. Only at an expansion surpassing the well known nucleon density of stable nuclei, will we have the formation of nuclei with hadron matter in their interior. In the reverse case, the generation of all kinds of nuclei may occur if a hadron ensemble is compressed from very low densities. Surpassing the well known stable nuclear density, the nuclei will be formed by the surface energy. A dependence on the nucleon number shows that nuclei with a number of nucleons above a limit (about 400) cannot be held together by the above plasma-Fermi surface tension, excluding the possibilities of obtaining superheavy nuclei.

We see how a very applied classical physics field such as laser produced plasmas can be generalized to describe such phenomena as the phase transition of hadron to quark matter when forming nuclei, understood from surface tension in agreement with the Yukawa potentials.

Acknowledgements

I am very grateful to Dr C S Taylor (CERN) for valuable comments when reading this text. Support by Dr Kurt Hubner, Dr Helmut Haseroth and Dr E J N Wilson (CERN) is gratefully acknowledged.

References

[1] J.H. Nuckolls. "Physics Today", 35 (No 9), 24 (1982); E. Strom et al LLNL Report No 47312, August 1988
[2] H. Hora. "Plasmas at High Temperatures and Densities", (Springer, Heidelberg 1991)
[3] C. Yamanaka, S. Nakai, T. Yamanaka, Y. Izawa, K. Mima, K. Nishihara, Y. Kato, T. Mochizuki, M. Yamanaka, M. Nakatsuka and T. Yabe. "Pellet Implosion and Interaction Studies by GEKKO XII Green Laser", Laser Interaction and Related Plasma Phenomena, H. Hora and G. H. Miley eds. (Plenum, New York 1986) vol 7, p395

[4] H. Hora, D. Pfirsch, and A. Schluter. "Z. Naturforsch" 22A, 278 (1967)

[5] J. W. Shearer. R. E. Kidder and J. W. Zink, " Bull. Amer. Phys. Soc." 15, 1483 (1970)

[6] H. Hora. "Phys. Fluids" 12, 182 (1969)

[7] H. Hora. "Phys. Fluids" 28, 3706 (1985)

[8] B. W. Boreham and H. Hora. "Phys. Rev. Letters" 42, 776 (1979)

[9] S. Augst, D. Strickland, D. D. Mayerhofer, S. L. Chin and J. H. Eberly. "Phys. Rev. Letters" 66, 1247 (1991)

[10] H. Hora and P. H. Handel. "New Experiments and Theoretical Development of the Quantum Modulation of Electrons (Schwarz-Hora Effect)" in Advances in Electronics and Electron Physics, P. W. Hawkes ed. (Acad. Press, New York, 1987) vol 69, p55

[11] H. Hora, P. Lalousis and S. Eliezer. "Phys. Rev. Letters" 53, 1659 (1984)

[12] S. Eliezer and H. Hora. "Phys. Repts." 172, 339 (1989)

[13] H. Hora. "Plasma Model for Surface Tension of Nuclei and the Phase Transition to the Quark Plasma" CERN-PS/DL-Note-91/05, August 1991

ROBERT DAUTRAY
1993

ACCEPTANCE of the EDWARD TELLER MEDAL[1]

Robert Dautray, Haut Commissaire à l'Énergie Atomique
Commissariat à l'Énergie Atomique
31-33 Rue de la Fédération
75752 Paris Cedex 15 - FRANCE

26 October 1993, Monterey, California

Dear Colleagues, I am greatly honoured and happy that you, gentlemen, granted me the Edward Teller Medal Award.

For many reasons, it is a great event in my life, as it means to me the culmination of lifelong work and endeavours to understand thermonuclear fusion when it is in bulk material, like in the stars, and to trigger it.

At the same time, I consider the name of this reward, *Edward Teller*, as a symbol of the science of our time, going from fundamental physics to applied sciences, from the laboratory to the welfare of our countries.

Which is more, this reward is coming from the United States of America, the country of freedom and democracy, who crossed oceans to free us, French people, from barbarians' strength and helped us out of the ruins of the most destructive war.

My personal gratefulness comes also from the fact that I keep a warm and cheerful remembrance of a lifelong collaboration with you, my dear colleagues of Livermore, Los Alamos, Rochester, Washington, Boston, whom I consider as my friends, hoping that is a reciprocal feeling.

Let us go on working together to arrive at the next step, ignition of a plasma in the laboratory.

As we say in France, *"un grand merci."*

---

[1] Due to the heavy demands of his new position in the French government, Dr. Dautray was was unable to attend the 11th Workshop. He was represented by his colleague, Dr. Jean-Paul Watteau, who read this acceptance speech on behalf of Dr. Dautray.

JOHN D. LINDL
1993

THE EDWARD TELLER MEDAL LECTURE: THE EVOLUTION TOWARD INDIRECT DRIVE AND TWO DECADES OF PROGRESS TOWARD ICF IGNITION AND BURN

John D. Lindl
Lawrence Livermore National Laboratory
Livermore, California 94550 USA

Eleventh International Workshop on Laser Interaction
Monterey, California

In 1972, I joined the Livermore ICF Theory and Target Design group led by John Nuckolls, shortly after publication of John's seminal Nature article on ICF.[1] My primary role, working with others in the target design program including Mordy Rosen, Steve Haan, and Larry Suter, has been as a target designer and theorist who utilized the LASNEX code to perform numerical experiments, which along with analysis of laboratory and underground thermonuclear experiments allowed me to develop a series of models and physical insights which have been used to set the direction and priorities of the Livermore program.

I have had the good fortune of working with an outstanding team of scientists who have established LLNL as the premier ICF laboratory in the world. John Emmett and the LLNL Laser Science team were responsible for developing a series of lasers from Janus to Nova which have given LLNL unequaled facilities. George Zimmerman and the LASNEX group developed the numerical models essential for projecting future performance and requirements as well as for designing and analyzing the experiments. Bill Kruer, Bruce Langdon and others in the plasma theory group developed the fundamental understanding of laser plasma interactions which have played such an important role in ICF. And a series of experiment program leaders including Mike Campbell and Joe Kilkenny and their laser experimental teams developed the experimental techniques and diagnostic capabilities which have allowed us to conduct increasingly complex and sophisticated experiments.

636   The Evolution toward Indirect Drive

Since 1975, for reasons described below, the Livermore ICF Program has
concentrated most of its effort on the Indirect Drive approach to ICF. In this
approach, the driver energy, from laser beams or ion beams, is first absorbed within
a high-z enclosure, a hohlraum, which surrounds the capsule. The material heated
by the driver emits x-rays which drive the capsule implosion. Typically, 70–80% of
the driver energy can be converted to x-rays. The optimal hohlraum geometry
depends on the driver. Schematic hohlraums for a laser and heavy ion beam driver
are shown in Fig. 1.

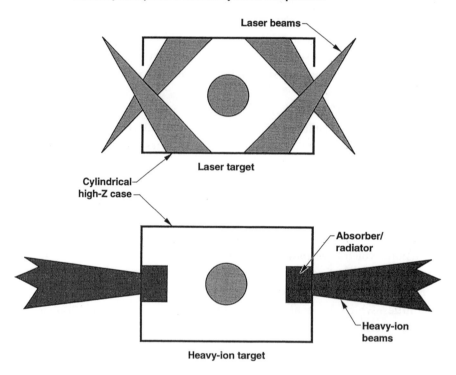

Figure 1. The optimal hohlraum geometry for indirect drive depends on the driver.

John D. Lindl   637

The target physics constraints on laser driven indirect drive ignition targets have been identified as shown in Fig. 2. Driver target coupling issues, which include laser absorption, x-ray conversion and transport, limit laser driven hohlraums to an x-ray intensity of about 10^{15} W/cm^2. This is primarily the result of laser driven parametric instabilities which result in scattering and production of high energy electrons which cause capsule preheat. Given the x-ray flux limitations, and the implosions velocity required for ignition, the flux onto the capsules must be sufficiently uniform to allow the capsules to converge a factor of 25–35. This convergence is defined as the ratio of the initial outer radius of the capsule to the compressed radius of the hot spot which ignites the fuel. For a capsule, to achieve convergences this large and remain nearly spherical, x-ray fluxes must be uniform to 1–2%. This kind of uniformity in hohlraums requires a hohlraum which is large compared to the capsule. Hohlraum areas are typically 15–25 times that of the initial capsule area. Such a large area hohlraum limits hohlraum coupling efficiencies to 10–15%. It should ultimately be possible to achieve a coupling efficiency of 20–25% through the use of optimal driver and hohlraum geometries.

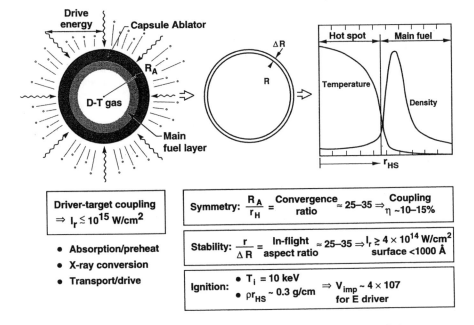

Figure 2. The Target Physics constraints on ICF ignition targets have been identified.

638   The Evolution toward Indirect Drive

Hydrodynamic instability places a limit of 25–35 on the ratio of shell radius R to shell thickness ΔR, as it implodes. As discussed below, the achievable implosion velocity, which is the primary determinant of the minimum size driver for ignition, is determined by a combination of the allowable capsule aspect ratio and the maximum achievable intensity.

Based on these constraints, the U.S. Department of Energy is considering building a National Ignition Facility (NIF) based on the Neodymium glass laser. This facility would be capable of delivering 1.8 MJ of 0.35 mm laser light with a peak power of 500 TW. The gains calculated for this facility are shown in Fig. 3.

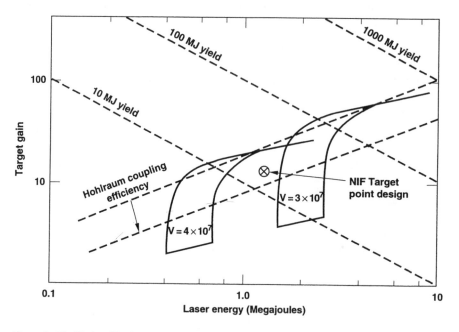

Figure 3. The National Ignition Facility (NIF), based on a neodymium glass laser, is being designed to demonstrate ICF capsule ignition and burn propagation.

John D. Lindl 639

We are nearing completion of a two decade quest to develop the data and numerical modeling capabilities needed to accurately specify the requirements for ignition and high gain radiation driven ICF targets. Some of the highlights of this effort are listed in Fig. 4. Throughout its history, research on indirect drive in the U.S. has been largely classified. A proposed DOE declassification should go a long way toward achieving a long time goal of Edward Teller and most of the research scientists in the field for more openness and international cooperation in ICF.

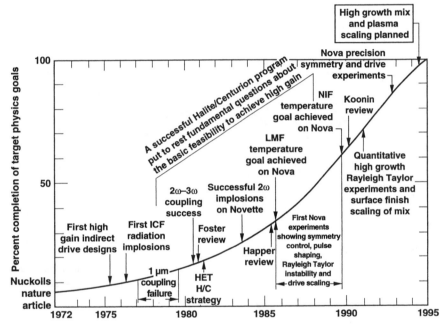

Figure 4. We are nearing completion of a two decade effort to develop the data and numerical modeling capability required to accurately specify the requirements for ignition and high gain ICF targets.

I have started the chronology in Fig. 4 with the publication of Nuckolls Nature paper, partly because that is when I joined the LLNL ICF Program, but also because it occurred approximately at the time when laser technology, diagnostic development and numerical modeling were becoming sufficiently mature that an expanded ICF program could begin to evaluate the limits and requirements for success of ICF. Nuckolls paper was based on the direct drive implosion of bare

640   The Evolution toward Indirect Drive

drops or shells of DT. But as indicated, by 1975 LLNL had shifted to radiation driven implosions. Before following the chronology in Fig. 4 further, I want to describe the series of developments which led to this change in direction.

Figure 5 shows Nuckolls original gain curves as a function of fuel compression. These curves predict that targets driven by lasers as small as 1 kJ could achieve target gains greater than unity if sufficient compression is achieved. Drivers of about 1 MJ were predicted to be required for high gain. Although today's estimates of the driver size required for high gain have not changed significantly, we now believe that about a 1 MJ driver will be required to achieve ignition as well.

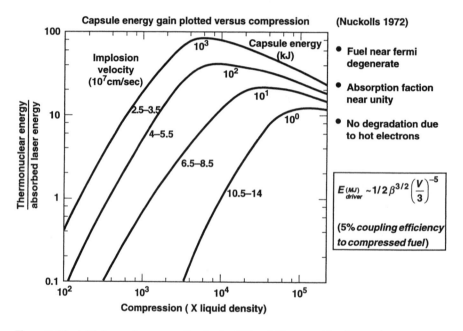

Figure 5. The initial capsule energy estimates for ICF could be met with a factor of 3-4 increase in the achievable implosion velocity compared to velocities predicted for the NIF.

Why has the ignition threshold increased by such a large factor while the requirements for high gain have remained fairly constant? This apparent disparity is explained by the extreme sensitivity of the ignition threshold to the implosion velocity which can be achieved in near spherical implosions in which the fuel remains nearly Fermi-degenerate. A simple model for an isobaric implosion[2]

John D. Lindl 641

would predict that the required energy would scale as $\beta^3 V^{-10}$ where V is the implosion velocity and β is the ratio of the pressure in the fuel to the Fermi pressure. Detailed numerical calculations predict that this dependency is approximately reduced to:

$$E_{capsule}(MJ) = 1/2(\frac{0.05}{\eta_{hydro}})\beta^{3/2}(V/3)^{-5} \tag{1}$$

where η_{hydro} is the hydrodynamic efficiency and V is in units of 10^7 cm/sec. The reduction in the velocity dependence from an ideal isobaric model is the result of several factors including an increase in the required hot spot temperature as the target size decreases, and an increase in the fractional mass in the hot spot which self consistently occurs during the compression and results in a reduced compression and hot spot coupling efficiency.

The strong dependence of the minimum energy on the achievable implosion velocity means that there is only a factor of about 4 in the implosion velocity required to go from a minimum energy of about 1 MJ to a minimum energy of 1 kJ, as shown in Fig. 5. The range of velocities indicated depends on the ignition margin required to overcome the effects of mix during the process of assembling the hot spot.

The dependence of gain on implosion velocity is much weaker. To zero order, the reduction in gain with reduction in target size, or increase in implosion velocity, occurs primarily because less mass is imploded per joule of energy coupled to the target. For equal burn efficiency, the gain would scale as V^{-2} so that there would be about an order of magnitude less gain at 1 kJ than at 1 MJ, as shown in the curves of Nuckolls. Once targets with a DT pusher ignite, they are expected to burn as calculated. In practice, compressions achieved in self consistent implosions are lower than the optimal indicated in Fig. 4 and the gain is a stronger function energy.

The targets being designed for the proposed NIF require velocities of $3\text{-}4 \times 10^7$ cm/sec. We now believe that implosion velocities are limited to values near this range because of the intimate coupling between the implosion velocity and the physics governing the hydrodynamic instability of an ICF implosion and the maximum intensity allowed by efficient laser plasma coupling. Higher implosion

642   The Evolution toward Indirect Drive

velocities are possible in certain types of high entropy implosions such as exploding pushers capsules but these targets do not scale to high gain.

The dispersion relation for the Rayleigh-Taylor instability used by Nuckolls in 1972 was one attributed to Leith based on a model of fire polishing during the ablation process:

$$\gamma^2 = ka - k^2 \frac{P_a}{\rho} = ka(1 - k\Delta R) \tag{2}$$

where P_a is the ablation pressure, k is the wave number, and ΔR is the shell thickness. This dispersion relation predicts that all wavelengths shorter than $2\pi\Delta R$ are stabilized and that the maximum number of e-foldings is approximately:

$$n_{max} = \int \gamma_{max} dt \approx \frac{1}{2}\sqrt{\frac{R}{\Delta R}} \tag{3}$$

for constant acceleration and $1/2at^2 = R/2$. The shell aspect ratio $R/\Delta R$ can be related to the implosion velocity, the shell adiabat and the laser intensity and laser wavelength by integrating the rocket equation for a directly driven laser implosion[3]:

$$\frac{R}{\Delta R} = 0.7 \frac{V^2}{V_{ex}V_{abl}} = \frac{60(V/3)^2}{\beta^{\frac{3}{5}}(I_{15}/\lambda)^{\frac{4}{15}}} \tag{4}$$

where I_{15} is the laser intensity in units of 10^{15} W/cm^2 and λ is the laser wavelength in microns. $V_A \equiv \dot{m}/\rho$ is the ablation velocity, $\dot{m}$ is the mass ablation rate and ρ is the peak density in the shell. V_{ex} is the exhaust velocity from the rocket equation:

$$V = -\frac{P}{\dot{m}} \ell n(m/m_0) \equiv -V_{ex}\ell n(m/m_0) \tag{5}$$

John D. Lindl   643

where m_0 is the initial shell mass and m is the final shell mass. P is the ablation pressure driving the implosion. Using the definitions of V_{ex} and V_{abl} , Eq. (4) can also be written:

$$\frac{R}{\Delta R} \approx 3/4 \frac{\rho V^2}{P}$$

In this form it is clear that the aspect ratio can be simply obtained by setting the energy in a shell of thickness ΔR equal to the PdV work that can done during an implosion over half the total volume of the shell.

Equations (1), (3), and (4) can be combined to obtain:

$$E_{capsule}(MJ) = 1/2\beta^{3/2}(V/3)^{-5} \approx 1/2(\frac{n_{max}}{10/3})^{-5} I_{15}^{-2/3} \tag{6}$$

where a typical direct drive coupling efficiency of 5% into the compressed fuel is assumed, very little margin is allowed for mix during assembly of the hot spot, and $\lambda=1/3$ μm is chosen for the laser wavelength. This transforms the minimum driver energy for direct drive into a relationship which depends only on the maximum allowed growth of hydrodynamic instabilities and the allowed laser intensity. If we let the maximum number of e-foldings equal 6, we get:

$$E_{capsule}(KJ) = 26 I_{15}^{-2/3} \tag{7}$$

If we allow a maximum intensity of 10^{17} W/cm^2, then the minimum driver energy is 1 kJ. In 1972, computers and numerical models were just becoming powerful enough to begin doing detailed evaluation of the effects of hydrodynamic instability and the limits that laser plasma instabilities would place on the allowable intensity. Over the next few years we learned that laser intensities would be limited to 10^{14} to a few times 10^{15} W/cm^2 depending on the laser wavelength. Numerical calculations provided most of the guidance for the growth of Rayleigh-Taylor instability until quantitative data became available in the late 1980's and the early 1990's.

By 1974, numerical calculations by Bill Mead and me,[4] using the LASNEX code, indicated that direct drive capsules would have much higher instability growth rates than indicated by Eq. (2). In addition, experiments had begun using

644   The Evolution toward Indirect Drive

Neodymium glass lasers which indicated that reduced absorption and hot electron production would severely degrade direct drive implosions at the high intensities required for ignition with lasers in the 1–100 kJ range. The laser beam quality was also much worse than could be tolerated for the implosion uniformity required for direct drive.

In hind sight, it is possible to write down a set of equations which quantify the limitations imposed on direct drive by these constraints. In 1985, Takabe et al.,[5] wrote down a dispersion relation for direct drive Rayleigh-Taylor instability. Takabe's formula

$$\gamma = 0.9\sqrt{ka} - \beta k V_A \qquad \text{where } \beta\text{=3-4} \tag{8}$$

has a far weaker form of ablation stabilization than that given by Eq. (2). Takabe's formula is obtained from a series of numerical calculations with very small density gradients at the ablation front. This formula can be modified[6] to account for density gradients which can result in significant stabilization in certain types of direct drive implosions:

$$\gamma = \sqrt{\frac{ka}{1+kL}} - 3kV_A \tag{9}$$

where L is the density gradient scale length and β=3 has been used. This density gradient stabilization correction was first proposed by Lelevier[7] based on a suggestion by Teller. Experiments[8] have now been carried out which agree with the numerical simulations and are in substantial agreement with Eq. (9) although the coefficients multiplying L and V_A are not universal. These experiments, having a variety of wavelength, initial amplitudes, and laser beam quality, have been carried out using two beams of the Nova laser as shown schematically in Fig. 6. Shown in Fig. 7 are the amplitude versus time for a variety of wavelengths with an initial amplitude of 1 μm. Also shown are the amplitudes versus time of the various harmonics which grow up as the shape of applied perturbation is distorted by nonlinear effects.

John D. Lindl   645

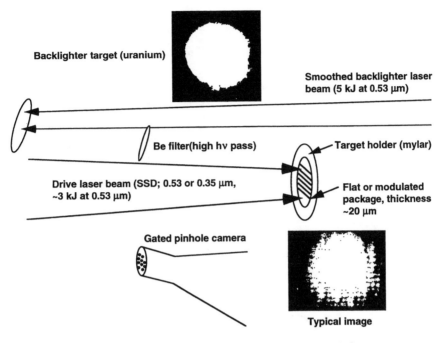

Figure 6. Planar direct drive Rayleigh-Taylor experiments have been carried out on
the Nova laser.

Integrating Eq. (8), the maximum number of e-foldings is given by:

$$n = \int \gamma dt = \sqrt{\frac{l}{1 + 0.1l\frac{\Delta R}{R}}} - 3l\frac{\Delta R}{R}(1 - m/m_0) \qquad (10)$$

where $\ell = k/R$ is Legendre polynomial mode number. A value for the density
gradient scale length L=0.1ΔR has been chosen. This is near the maximum that has
been achieved in calculations of direct drive implosions designed to minimize
Rayleigh-Taylor instability. The quantity m/m_0 comes from the rocket equation,
Eq. (5) above.

646   The Evolution toward Indirect Drive

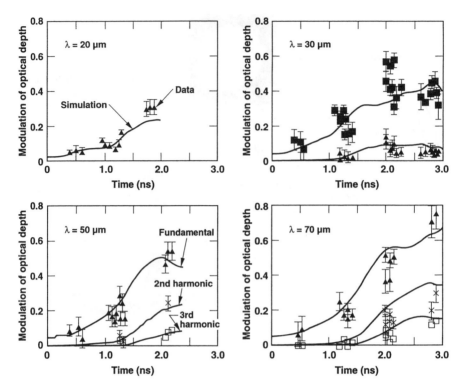

Figure 7. Experimental data agree with predictions at all harmonics and wavelengths studied ($a_0 = 1.0$ μm).

For direct drive, over the range of velocities and intensities of interest to ICF,

$$(1 - m/m_0) \approx 0.2(V/3)I_{15}^{-0.3} \tag{11}$$

Using Eqs. (4) and (11) in Eq. (10), the maximum number of e-foldings is then given approximately by:

$$n_{max} \approx 8.5\beta^{-2/5}(V/3)^{1.4}I_{15}^{-1/15} \tag{12}$$

again choosing λ=1/3 μm as the laser wavelength. Combining this equation with Eq. (1) for a 5% efficient coupling between driver and imploding fuel, the required driver energy becomes:

John D. Lindl   647

$$E_{driver}(MJ) \approx 1/2(n_{max}/8.5)^{-25/7} I_{15}^{-1/4} \qquad (13)$$

Eq. (13) has a much higher ignition threshold, and a much weaker dependence on laser intensity than Eq. (7). If we allow a maximum amplification of 1000 (6.9 e-foldings) then we have:

$$E_{driver}(MJ) \approx 1.0 I_{15}^{-1/4} \qquad (14)$$

This result, which allows little margin for degradation due to mix or asymmetry during assembly of the hot spot, is consistent with recent calculations[9] at the University of Rochester as shown in Fig. 8.

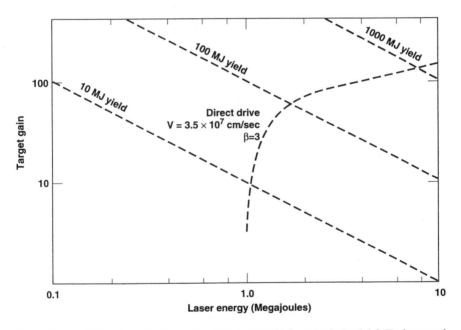

Figure 8. Calculations from the University of Rochester which constrain Rayleigh-Taylor growth to 6-7 e-foldings predict about a 1 MJ threshold laser energy for direct drive.

A great deal of progress has been made toward solving the irradiation uniformity problem for direct drive. A series of clever optical inventions both in the U.S. and Japan[10-12] has traded off laser beam coherence for laser beam uniformity

648   The Evolution toward Indirect Drive

so that it is now possible to obtain beams which are uniform to a few percent. The overlap of a large number of beams and further optical innovation should allow the achievement of the <1/2% nonuniformity estimated to be required for direct drive. Further optimization of density gradient stabilization effects may allow for a reduction in the direct drive threshold below the 1 MJ estimated above. In the U. S., the Omega Upgrade laser under construction at the University of Rochester is designed to answer these questions, and the NIF conceptual design will be consistent with the addition of a direct drive option. In Japan, the Gekko XII laser is being upgraded for the same purposes.

After the 1974 direct drive hydrodynamic instability calculations, I began an effort to develop high gain radiation implosions. A primary virtue of radiation driven implosions is the fact that they have much higher ablation rates and thicker shells which make control of the Rayleigh-Taylor instability more feasible. In addition, because the laser beams are absorbed far from the capsule, as shown in Fig. 1, radiation driven implosions are unaffected by small scale laser beam nonuniformities. Also because of solid angle effects, and the fact that radiation driven capsules are thicker than directly driven capsules, indirect drive targets are less sensitive to the effects of hot electrons produced by laser driven parametric instabilities. A disadvantage of indirect drive for lasers, compared to direct drive, is the longer scalelength of plasma traversed by the laser as it propagates from the laser entrance hole to the hohlraum wall.

For indirect drive, the capsules and such issues as radiation transport, and hohlraum wall loss are essentially independent of the driver. Hence many of the results learned with lasers carry over to other drivers such as heavy ion beams. This synergism is particularly important for heavy ion beams drivers. Because of their efficiency, durability and high rep rate capability, these drivers are considered prime candidates for an Inertial Fusion Energy application. However, the research required to define the characteristics of a heavy ion beam driver that could reach the beam intensities required for ICF experiments is still incomplete. Indirect drive laser experiments provide a key element of the data base required to insure that heavy ion beam driven targets will work when the accelerators are available.

In 1975, I showed that it is possible, with proper design, to achieve high gain with radiation driven targets such as those in Fig 1. These targets are very similar to

John D. Lindl   649

direct drive targets, except that the choice of ablator material must be properly matched to the x-ray drive spectrum in order to insure that the fuel can be kept in a near Fermi degenerate state. From one point of view, soft x-ray driven implosions could be considered as being driven by a very short wavelength, very broad band laser. LASNEX calculations in 1975 showed a dramatic reduction in growth compared to direct drive. The dispersion relation which I obtained from these simulations is:

$$\gamma = \sqrt{\frac{ka}{1+kL}} - kV_A \tag{15}$$

and

$$n = \int \gamma dt \approx \sqrt{\frac{l}{1+0.1l\frac{\Delta R}{R}}} - 0.8l\frac{\Delta R}{R} \tag{16}$$

Eq. (15) which I wrote down in 1982 is essentially the same as for direct drive, except the kV_A term is multiplied by 1 instead of 3. Even with the lower coefficient multiplying the ablation term, the much higher ablation rates and the resultant thicker shells and larger density scale length result in lower growth for indirect drive under conditions required for ICF. Ablation stabilization can be understood as a simple convective effect due to the fact that Rayleigh-Taylor modes are surface modes. In the absence of ablation, the amplitude η of a Rayleigh-Taylor mode is given by:

$$\eta = e^{\gamma_0 t}e^{-kx_0} \tag{17}$$

where γ_0 is the growth rate in the absence of ablation and x_0 is position measured relative to the initial surface of the shell. In an ablating shell, the only material that matters to the implosion is the material inside of the ablation front. The ablated material disappears beyond the sonic horizon of the shell and no longer affects the implosion. When distance into the shell is referenced relative to the ablation front X_A, Eq. (17) becomes:

$$\eta = e^{\gamma_0 t}e^{-(kx_0 + kV_A t)} = e^{(\gamma_0 - kV_A)t}e^{-kx_A} \tag{18}$$

650   The Evolution toward Indirect Drive

and there is an effective reduction in growth. In this model with a uniform density shell, it makes sense to define the ablations velocity relative to the peak density, as was done earlier. However, in an actual experiment, the Rayleigh-Taylor mode will not necessarily be located near peak density. The coefficient multiplying V_A varies by a factor of about 2 in numerical simulations and this could be due to variations in the position of the modes relative to the peak density. The factor of 2–3 difference between direct drive and indirect drive could also be due to this effect.

Nova experiments[13] and calculations for indirect drive Rayleigh-Taylor instability are in substantial agreement and are consistent with Eq. (15). Figure 9 shows a typical setup for these experiments. Data are obtained by looking at an x-ray backlighter through a sample placed near the hohlraum wall. By looking face-on through a perturbed sample using either an x-ray framing camera or an x-ray streak camera, the flow of material from thin to thick regions can be observed as an increase in the contrast in x-ray transmission through the sample. Representative data and calculations are shown in the upper right hand corner of Fig. 9. By looking edge-on to a perturbed sample, the shape of the perturbations can be recorded by using an x-ray framing camera. Representative data and calculations are shown as the middle two images in Fig. 9. By using an x-ray streak camera and looking edge-on, the sample position versus time can be recorded. Representative data and calculations are shown in the bottom two images in Fig. 9.

By integrating the rocket equation for indirect drive, we can obtain a relation between the shell aspect ratio, the implosion velocity and the ablation velocity:

$$V/3 \approx 1/2 \frac{R}{\Delta R} \frac{V_A}{3} \approx \frac{R/\Delta R}{22} \beta^{3/5} I_{15}^{9/40} \tag{19}$$

If we limit the maximum number of e-foldings to about 6, then Eq. (16) implies that $R/\Delta R = 30$. Using Eq. (1) with a 15% hydro efficiency we obtain for the capsule energy:

$$E_{capsule}(KJ) \approx 35 \beta^{-3/2} I_{15}^{-9}/8 \tag{20}$$

John D. Lindl   651

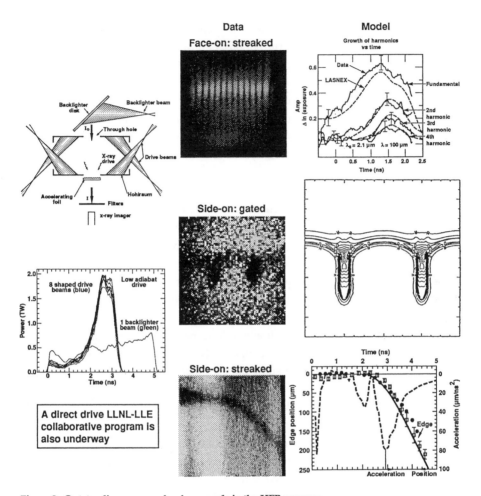

Figure 9. Outstanding progress has been made in the HEP program.

652   The Evolution toward Indirect Drive

The maximum tolerable capsule aspect ratio depends on the capsule surface finish and the spectral distribution of perturbations. Central to the ICF strategy for igniting the minimum size capsule is keeping growth small enough that these hydrodynamic instabilities remain in the linear or only weakly non-linear regime.[14]   In this regime, mode coupling is small and the performance of capsules can be predicted by taking an RMS sum of the effects of the whole spectrum of initial perturbations either initially present on a capsule or impressed during the implosion process by such effects as intensity non-uniformity. All of these considerations result in an In-flight capsule aspect ratio of about 30.

Equation (20) gives capsule energies very close to the projections in Nuckolls 1972 paper, and in fact scales more favorably with intensity. However, there is a price to be paid in hohlraums to produce x-rays, and to transport them symmetrically to a capsule. Further, there are limits to the achievable x-ray intensity. With these considerations, the minimum driver energy for indirect drive is also approximately 0.5–1 MJ. The larger energy allows for the effects of hot and cold fuel mixing together while the hot spot is being assembled. Hence the currently projected minimum energy for ignition and burn propagation are quite similar for both direct and indirect drive. The proposed National Ignition Facility is being designed to be consistent with pursuing both approaches.

However, because of relaxed beam quality requirements and reduced sensitivity to Rayleigh-Taylor instability, it has been possible over the past two decades to make more rapid progress toward obtaining the radiation drive data base required to quantitatively specify the driver requirements for ignition. In the U.S., a comparable data base for direct drive will not be available until completion of the experiments planned for the Omega Upgrade, some time after the end of the decade.

In 1976, I designed targets for the first laser driven radiation implosions. Targets based on these designs, which required about 100 joules of 1 μm laser light, were fielded on the Cyclops laser at LLNL in the spring of 1976 and worked essentially as calculated. These results made us extremely optimistic about radiation drive, resulted in a shift of the Shiva laser from a uniform illumination scheme to a two sided irradiation scheme for direct drive, and formed the basis for the early optimism about the possibilities for achieving breakeven on a 200 kJ twenty beam Nova laser.

John D. Lindl   653

However, as we tried to achieve higher drive temperatures for high density implosions and higher yields, parametric instabilities generated high levels of energetic electrons. Coupling difficulties hampered progress in the LLNL program on both the Argus and Shiva laser as we worked to achieve a 100X liquid DT fuel density implosion goal.

In the initial Cyclops hohlraum experiments, the presence of high energy electrons, which we later determined were produced by Stimulated Raman Scattering, showed up as noise in the neutron detectors. We tested three different size hohlraums. The smaller two had such a large high energy noise signal that they could not be used for implosions. So we did the implosion experiments in the larger lower temperature hohlraum which had low noise levels. We were later to find that laser plasma parametric instabilities strongly limit the hohlraum temperature that can be achieved with a given size and wavelength laser.

Because we lacked adequate models for certain key pieces of the physics, including NLTE atomic physics for the high-z hohlraum walls, and adequate understanding of the laser plasma interactions below critical density inside the hohlraum, we believed that absorption would be high in hohlraums for 1 μm light and that conversion to x-rays would be very high. We fully expected to achieve 100X liquid density on the Argus laser and be able to go on to more aggressive implosions on Shiva. We did achieve implosions near 100X on Shiva but primarily spent the years on Shiva and Argus developing an understanding of the source and limitations imposed by hot electron production in hohlraums.

In 1978, based on an examination of the plasma conditions calculated for the Shiva hohlraums, and an improved understanding of Stimulated Raman Scattering developed by Bill Kruer and Bruce Langdon, I proposed a model for the limitations on hohlraum temperatures as a function of laser wavelength. This model predicted that we would be able to achieve the temperature required for high gain if the proposed Nova laser were built to produce the third harmonic of the neodymium glass laser by using non-linear conversion in KDP crystals. This became practical because of efficient conversion schemes initially devised by the University of Rochester.[15] The improved coupling in hohlraums was demonstrated at the 100 J level in a series of experiments by Mike Campbell on the Argus laser in 1980-81.

654   The Evolution toward Indirect Drive

In 1979, I used these estimates of the achievable radiation temperatures and an estimate of the hohlraum coupling efficiency which could be achieved consistent with the required implosion symmetry to estimate the single shell target gains shown in Fig. 10. These gains were very similar to those now estimated for the National Ignition Facility.

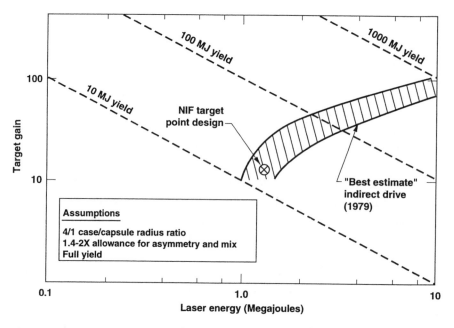

Figure 10. By 1979 "best estimate" gains for single shell radiation driven targets were comparable to today's estimates.

In 1981, when it became clear that ignition would not be achieved on the Nova laser, I devised a strategy for obtaining the data base which would be required for ignition on a future facility. This strategy tests the physics of high gain targets by using a series of Nova experiments on targets which are as close as possible to being "Hydrodynamically Equivalent Targets" (HET) and a series of underground experiments at much larger energy. This series of experiments would be the ICF equivalent of wind tunnel tests to provide the basis for the facility that would ultimately be required for ignition and high gain.

John D. Lindl 655

Since 1981, the LLNL ICF program has focused on carrying out the work needed to develop the quantitative modeling tools, diagnostics, and experimental techniques required to develop this physics basis. This strategy was endorsed by the Foster Review in 1981, along with the recommendation that Nova be a smaller 10 beam laser with 2ω and 3ω capability. These recommendations were implemented and Nova has proved to be an extremely successful facility. In addition, the Halite/Centurion program, which was a joint Livermore/Los Alamos program of underground experiments, put to rest fundamental questions about the basic feasibility of high gain.

When John Nuckolls left the ICF Program in 1983 to assume leadership of the LLNL Physics department, I assumed responsibility for the ICF Theory and Target Design Program. In that year, the first two beams of Nova were constructed as a laser test bed and experimental facility called Novette which gave us our first multikilojoule experience with 2ω light. ICF experiments on this facility only lasted about three months but indicated that we were very likely to achieve our hohlraum temperature goals on Nova.

The Happer review by the NAS occurred just as Nova was being completed at the end of 1985 and little data was available. This review continued to endorse the HET approach but made the Halite/Centurion program the highest priority.

Between 1986 and 1990, LLNL made rapid progress under the ICF Program leadership of Erik Storm. During this period, the Experimental Program was led by Mike Campbell and I led the Theory and Target Design Program. Nova achieved its initial temperature goals soon after being activated. We had established this goal, based on my 1978 analysis, as the temperature which would be required for high gain with a 5–10 MJ laser. DOE labeled such a facility the Laboratory Microfusion Facility (LMF). Between 1986 and 1990, Nova experiments and quantitative modeling were carried out which showed symmetry control, the expected benefits of pulse shaping, the first quantitative Rayleigh-Taylor instability experiments, and the radiation drive temperature scaling of implosions.

Also, under the leadership of Hank Shay, Steve Haan, and Tom Bernat at LLNL and the leadership of Tom McDonald at LANL, the Halite/Centurion experiments rapidly achieved their goals.

656   The Evolution toward Indirect Drive

In 1990, I developed a model which quantified the dependence of the minimum driver energy on the achievable hohlraum temperature. Shortly after this, using the increased power and energy which had become available when Nova's laser glass was replaced with improved platinum free material, we were able to demonstrate the higher hohlraum temperatures that would be required for ignition and gain experiments with a 1–2 MJ laser. We proposed such a facility to DOE and the Koonin NAS committee which was in the process of reviewing the National ICF Program. Koonin's committee further refined the "hydrodynamic equivalence" effort by endorsing a series of specific goals in hohlraum and capsule physics which have become known as the Nova Technical Contract. Predicated on successful completion of these goals, this review recommended construction of a 1–2 MJ ignition facility based on a Neodymium glass laser as the next logical step for ICF.

The gains calculated for this facility, the NIF, are indicated in Fig. 3. Shown are gain curves at implosion velocities of 3 and 4×10^7 cm/sec, under the assumption of a fixed hohlraum coupling efficiency. At any given velocity, capsules below a certain energy, as discussed above, will fail to ignite because the hot spot will not achieve sufficient ρr and temperature. The shaded bands correspond to a minimum cutoff energy which depends on hydrodynamic instability levels and capsule surface quality. The left-hand edge of each band corresponds to the gain for perfectly uniform implosions. The right-hand edge of each band corresponds to the gain for targets with surface finishes of 500–100 $\overset{\circ}{A}$. On Nova, we have demonstrated hohlraum drive temperatures consistent with a velocity of 4×10^7 cm/sec.

In the year following the 1990 NAS review, I took on responsibility for the overall target physics program and Mike Campbell became the ICF program leader as we worked toward meeting the laser technology and target physics goals endorsed by the NAS. As part of its conclusions, the Koonin committee recommended that Los Alamos become an integral part of the Nova physics program and that work has been jointly supported by LLNL and LANL since the end of 1990. A key element of the Nova program from 1991 to 1993 has been an extensive series of hohlraum symmetry experiments. A large number of people from LLNL and LANL participated in these experiments which formed the first major collaborative series on Nova. Allan Hauer from LANL was the lead experimentalist and Larry Suter from LLNL was primarily responsible for the analysis. For large hohlraum to capsule size ratios, hohlraums effectively smooth all but the longest

John D. Lindl   657

wavelength perturbations. These long wavelengths must be eliminated by proper choice of laser or hohlraum geometry. As indicated in Fig. 11, the time integrated symmetry on Nova is controlled by varying the laser pointing or the hohlraum length. For an elongated hohlraum or for pointing which places the laser beams near the laser entrance hole, the average flux is pole high and the implosion is pancaked. As the beams are moved toward the midplane of the hohlraum, the flux becomes higher around the waist of the capsule which then implodes to a sausaged shape. In between these two extremes, the time average flux can be made uniform to better than 1%, as required for high gain capsules. Figure 12 shows the ratio of the two axes of the imploded capsule, as the pointing is varied, for a 3:1 contrast, 2.2 ns pulse. A variety of pulse shapes have been tested and the experiments and calculations are in quite close agreement across the entire data base. Late in 1992, we carried out the first implosions which utilized a series of improvements to Nova's power balance and pointing accuracy. These improvements have allowed us to achieve extremely reproducible implosions as shown in Fig. 13. The shot to shot symmetry variations are less then 1/2% RMS, consistent with the requirements for the time averaged symmetry required for ignition targets.

In early 1993, DOE endorsed the mission need for the National Ignition Facility and authorized development of a Conceptual Design Report (CDR) for the facility. The coming year is a critical period for the ICF Program as we work to complete the target physics and laser technology goals and the CDR. In the Target Physics Program, we have two primary goals. We are working to develop large size plasmas which mock up conditions in NIF scale hohlraums. Using these plasmas, we plan to quantify the laser capabilities which will be required to maintain efficient laser-plasma coupling at the hohlraum temperatures required for ignition on the NIF. In addition, we are working to complete a series of mix experiments with increasing hydrodynamic instability growth.

If all goes well, we hope to have DOE's endorsement for proceeding with the next step of the NIF and a recommendation to the Congress that the NIF be approved for a 1996 start. We would then be well on our way to completing the last leg toward demonstrating the feasibility of ICF. Demonstration of ignition and propagating burn would complete the physics basis for ICF and open up the path for applications of ICF including its utilization as a source for fusion power production.

658   The Evolution toward Indirect Drive

X-ray imaging of the imploded capsules confirm that standard Nova hohlraums provide symmetric radiation drive of the capsule

Standard Nova
hohlraums are
tuned to produce
symmetric drive

We use distortion (a/b)
as our measure of
image distortion in
detuned hohlraums

We have systematically detuned the symmetry by varying the laser beam pointing and monitored changes in the capsule image shapes to test our understanding of implosion symmetry

Outward pointing shifts   Symmetric   Inward pointing shifts
produce "pancake" distortion   produce "sausage" distortion

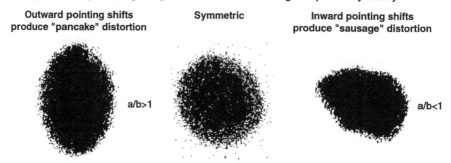

a/b>1   a/b<1

Observed X-ray image shapes confirm the calculated variation of the implosion shape with beam pointing*

Figure 11. We have used x-ray imaging of imploding capsules to test our modeling of hohlraum drive asymmetries.

John D. Lindl   659

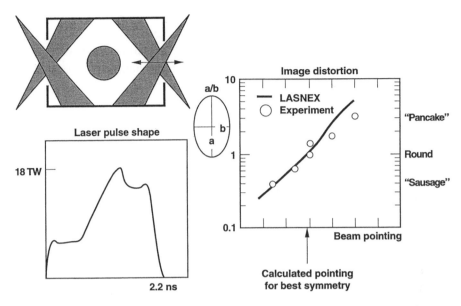

Figure 12. We are able to accurately model capsule distortion as a function of beam pointing.

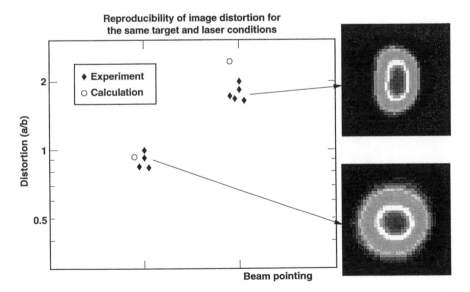

Figure 13. We have shown that our implosions are highly reproducible.

660   The Evolution toward Indirect Drive

ACKNOWLEDGMENT

Work performed under the auspices of the U.S. Department of Energy by the Lawrence Livermore National Laboratory under contract Number W-7405-ENG-48.

John D. Lindl 661

REFERENCES:

1. J.H. Nuckolls, L. Wood, A. Thiessen, and G.B. Zimmerman, "Laser Compression of Matter to Super-High Densities: Thermonuclear (CTR) Applications", Nature 239, 139 (1972)

2. Meyer-Ter-Vehn, Nucl. Fusion 16, 405 (1976)

3. J.D. Lindl, in "Fusion to Light Surfing", T. Katsouleas, ed., Addison-Wesley, Redwood City, Calif. (1991) p. 177

4. J.D. Lindl and W.C. Mead, "Two-Dimensional Simulation of Fluid Instability in Laser-Fusion Pellets", Phys. Rev. Lett. Vol. 34, Num. 20, (1975) p. 1273

5. H. Takabe, K. Mimo, L. Montierth, and R.L. Morse, Phys. Fluids 28, 3676 (1985)

6. M. Tabak, D.H. Munro, and J.D. Lindl, Phys. Fluids B 2, 1007 (1990)

7. R. Lelevier, G. Lasher, and F. Bjorklund, "Effect of a Density Gradient on Taylor Instability", National Technical Information Service Document No. DE86002577, National Technical Information Service, Springfield, Virginia, 22161 (1955)

8. S.G. Glendinning et al., "Laser Driven Planar Rayleigh-Taylor Instability Experiments, Phys. Rev. Lett. 69, 1201 (1992)

9. Unpublished presentation to the Inertial Confinement Fusion Advisory Committee, Aug 25–27, 1993

10. Y. Kato, K. Mima, N. Miyanaga, S. Arinaga, Y. Kitagawa, M. Nakatsuka, C. Yamanaka, Phys. Rev. Lett. 53, 1057 (1984)

11. S. Skupsky, R.W. Short, T. Kessler, R.S. Craxton, S. Letzring, and J.M. Soures, J. Appl. Phys. 66, 3456 (1989)

12. R. Lemburg, A. Schmitt, and S. Bodner, J. Appl. Phys, 62, 2680 (1987)

13. B.A. Remington, S.W. Haan, S.G. Glendinning, J.D. Kilkenny, D.H. Munro, R.J. Wallace, "Large Growth Rayleigh-Taylor Experiments Using Shaped Laser Pulses", Phys. Rev. Lett. 67, 3259 (1991)

14. S. Haan, "Weakly Nonlinear Hydrodynamic Instabilities in Inertial Fusion," Phys. Fluids B3, 2349 (1991)

15. W. Seka et al., Opt. Commun, 34, 469 (1980); R. S. Craxton, ibid., p. 474

SADAO NAKAI
1993

VIEWS on INERTIAL FUSION ENERGY DEVELOPMENT
Edward Teller Award Memorial Lecture
Monterey, California   26 October 1993

S. Nakai
ILE, Osaka University

It is my great honor to receive the Edward Teller Award. Representing the Institute of Laser Engineering, Osaka University, I would like to appreciate your favorite recognition on our achievements in laser fusion research.

Yesterday I talked the review of the activity and recent results of our Institute. This evening I would like to talk more personal feeling on the Inertial Fusion Energy development and surrounding conditions.

Figure 1 shows the world primary energy consumption as a function of year. I was surprised with the fact that the energy consumption increased so rapidly after the world war II and it was fought with this very low level of energy consumption. I clearly remember the miserable situation during and after the war even I was small child in elementary school. With this increased energy consumption, no more wars. Instead the war between nations, between peoples, we have common goal to collaborate, it is to keep our Earth sound for human beings.

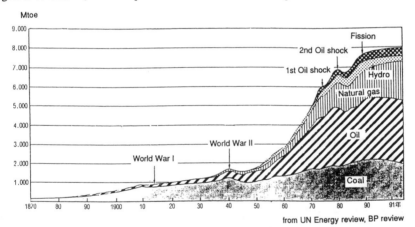

from UN Energy review, BP review

Fig. 1   World Primary Energy Consumption

The population of the world is clearly increasing so rapidly. The energy consumption per capita will also increase with the improvement of the quality of life through out the world. The increase of energy consumption in the past was supported mainly by the increase of fossil fuel consumption as shown in Fig. 1. What can support the increase of energy consumption in the future. Fig.2 shows the variation of CO_2 content in the atmosphere, one is the real measurement and the other for longer time span is the evaluated results from the glacier ice which was formed at respective age. It is clear that we can not rely on the massive consumption of fossil fuel in the future, due to the positive correlation between fossil fuel consumption and CO_2 concentration, and grovel worming by CO_2 which is still under investigation. So we are facing to a real war.

Sadao Nakai   663

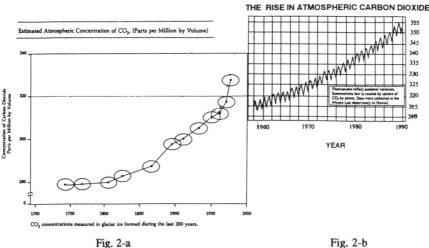

Fig. 2-a   Fig. 2-b

The energy sources such as hydro, solar, wind, tide and biomas are not abundant enough to support the civilized condition of human beings. Nuclear energy can support human activity keeping the earth to be sound for human existence. Fission is now popular but have some difficulties in public acceptance. Advanced nuclear technology must be developed to establish the generic nuclear technology and to increase the popularity and reliability of the nuclear energy. On this technology bases, we can utilize the fusion energy. Without the advanced generic nuclear technology we can not utilize even the fusion energy.

Recent progress of laser fusion research has urged us to initiate the world wide collaboration. So far, there had been many international collaborations on the bilateral bases. Following to the agreement at IAEA-Technical Committee Meeting in Osaka at 1991, we have been proceeding the world wide collaborative work under the framework of IAEA. It is entire review working of physics and technologies which is relevant to Inertial Fusion Energy, under the coordination by IAEA Advisory Group for IFE Review. Six advisory members and fifty experts through out the world have worked together for the IAEA Review Book "Energy from Inertial Confinement Fusion". It has almost completed and is now final editorial procedure for publication. Table 1 shows the major conclusion of the IAEA review work.

Table 1 The Major Conclusion of IAEA Review "Energy from Inertial Fusion"

(1)	Inertial Fusion ignition and energy gain can be achieved with laser technology now in hand and will probably be achieved in less than a decade.
(2)	Sufficient gain for a reactor appears to be achievable for drive energies of a few megajoules.
(3)	Credible technological concepts to carry out all the necessary function of an IFE reactor have been identified and it appears that a practical flexible IFE reactor can be built.
(4)	IFE reactors are likely to be economically acceptable, and can be designed to be safe and clean.
(5)	IFE development would benefit greatly by increased activity on technology development and increased international cooperation.

664   Inertial Fusion Energy Development

Now I would like to turn the topic. Fig. 3 is the famous drawing of Leonardo Da Vinci which is composed of edy lines. He knew the significance of Turbulence in nature. Turbulence and instability is one of the old physics but has still new meaning and significance in interpretation of the fluid physics in implosion fusion. by introducing the turbulent mixing model into the simulation code, our LHART (Large High Aspect Ratio Target) experiments were reproduced as shown in Fig. 4, where ξ is the ratio of assumed turbulent energy to the kinetic energy of implosion. At the aspect ratio where the stagnation is less dominant around 500, the inter penetration of SiO_2 and DT gas is less as be seen in Fig. 5. The detail analysis of the effect of turbulence in implosion are now very interesting topics.

Leonardo Da Vinci
(1452-1519)
Fig. 3-a

Fig. 3-b

Inertial Confinement Fusion is opening new fields of physics and advanced technologies. As an approach toward fusion energy, the feasibility is becoming more and more clear to believe. The IAEA review work on IFE is the first example of the world wide collaboration to do together even not a big project. It is anticipated that a real joint project like ITER in Magnetic fusion would be organized for the fusion energy development by Inertial Confinement .

Thank you for your attention.

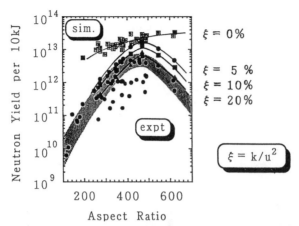

Fig. 4 1D simulations with the turbulent mixing
for ξ=10% reproduce the experiments

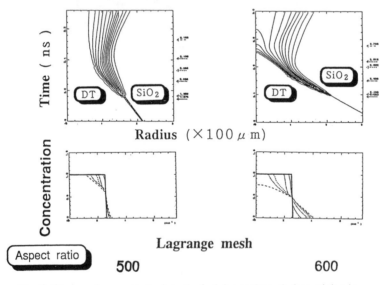

Fig. 5 Glass pusher penetrates into the fuel due to the turbulent mixing in
stagnation-dominant implosion, while the penetration is less for the
implosion with less stagnation.

E. MICHAEL CAMPBELL
1995
(read by Melissa Cray)

Path To Ignition:
US Indirect Target Physics

M. Cray[1] and E. M. Campbell[2]

[1]*Los Alamos National Laboratory, Los Alamos, New Mexico 87545 USA*
[2]*Lawrence Livermore National Laboratory, Livermore, California 94550*
USA

Abstract. The United States ICF Program has been pursuing an aggressive research program in preparation for an ignition demonstration on the National Ignition Facility. Los Alamos and Livermore laboratories have collaborated on resolving indirect drive target physics issues on the Nova laser at Livermore National Laboratory. This combined with detailed modeling of laser heated indirectly driven targets likely to achieve ignition, has provided the basis for planning for the NIF.

A detailed understanding of target physics, laser performance, and target fabrication is required for developing robust ignition targets. We have developed large-scale computational models to simulate complex physics which occurs in an indirectly driven target. For ignition, detailed understanding of hohlraum and implosion physics is required in order to control competing processes at the few percent level. From crucial experiments performed by Los Alamos and Livermore on the Nova laser, a comprehensive indirect drive database has been assembled. Time integrated and time dependent measurements of radiation drive and symmetry coupled with a detailed set of plasma instability measurements have confirmed our ability to predict hohlraum energetics. Implosion physics campaigns are focused on underdstanding detailed capsule hydrodynamics and instability growth. Target fabrication technology is also an active area of research at Los Alamos, Livermore, and General Atomics for NIF. NIF targets require developing technology in cryogenics and manufacturing in such areas as beryllium shell manufacture. Descriptions of our NIF target designs, experimental results, and fabrication technology supporting NIF target performance predictions will be given.

1. INTRODUCTION

The U.S. Inertial Confinement Fusion (ICF) Program is preparing to proceed with a next-generation ICF facility to demonstrate ignition: the National Ignition Facility (NIF). NIF is presently conceived as a Nd:glass laser facility with 192 beams capable of delivering 1.8 MJ of 351 nm laser light in a shaped pulse to an indirect-drive target. In indirect drive,[1] the energy of the laser or particle-beam driver is deposited inside an enclosure made of a high atomic-number material (hohlraum) which surrounds the fusion capsule. The hohlraum wall radiates X-rays which ablate the outer capsule wall, driving an implosion. NIF will also have a direct-drive capability using approximately 1.5 MJ. (In direct drive, the laser illuminates the capsule directly.) The NIF will be capable of 500 TW peak power, while maintaining a better than 50μm pointing accuracy and a less than 8% RMS beam to beam power imbalance over a 2 ns interval. The facility will be capable of handling up to a 45 MJ of D-T fusion yield.

Experiments at the Nova laser [2] by both LLNL and LANL personnel are resolving the key issues to ensure success of indirect drive on NIF. These issues are the physics related to capsule-illumination symmetry, laser-plasma instabilities and hydrodynamic instabilities.

With the LASNEX code, [3] state-of-the-art radiation-hydrodynamics modeling has been used to design NIF capsules and to understand to sensitivity of the designs on laser performance, capsule surface finish, hohlraum performance, etc.

The beamlet laser at LLNL is been used to establish the validity of the novel laser technology to be used in NIF. Another important technological area for NIF is target fabrication, particularly the manufacture of sufficiently smooth cryogenic D-T layers. Present capability has advanced to within a factor of two of the roughness presently deemed tolerable for ignition.

2. NIF TARGET DESIGN AND FABRICATION

Various NIF targets have been explored by designers from both LLNL and LANL[4]. All these capsules are imploded by the X rays from the same hohlraum, namely, a gas-filled Au cylindrical hohlraum with a length of 9.5 mm and a diameter of 5.5 mm (see Fig. 1). The hohlraum gas fill is kept by polyimide

($C_{22}H_{10}O_5N_2$) windows. Here we describe briefly two capsule designs for which LASNEX modeling indicates ignition. The modeling uses the integrated technique, where the laser, the hohlraum, and the capsule implosion and burn are all calculated self consistently in a single simulation.

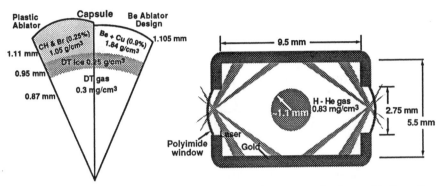

FIGURE 1. The ignition hohlraum for NIF is shown (right). Its size is roughly three times that of typical Nova hohlraums. On the left, details of two ignition capsules are shown.

The first target has a Br-doped CH outer ablator layer. We have used a radiation temperature profile T_r which reaches 300 eV to drive NIF capsules in our simulations (no mix of fuel and ablator due to hydrodynamic instabilities) yield for this target is 12MJ. The second ignition capsule has a Cu-doped outer ablator layer. LTE LASNEX modeling shows that with 1.30 MJ of laser energy (and 330 eV T_r), the clean fusion yield is 7 MJ. Other details of these targets are shown in Fig.1.

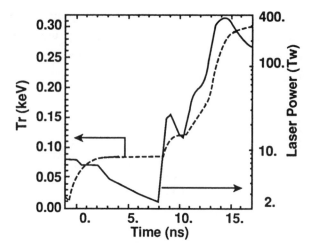

FIGURE 2. Idealized evolution of T_r and required laser power for the CH ignition capsule.

In order to achieve the capule compression and timing of the shocks which creates a hot spot in the capsule which ignites the propagating burn, various processes in hohlraum plasmas must be controlled. These include laser ray propagation, refraction, absorption and X-ray conversion; multi-group X-ray transport; hohlraum plasma dynamics (filling, stagnation, heat conduction), and; capsule evolution and burn. Successful capsule compression and shock timing to ignite a spark in the fuel that initiates the propagating burn requires a highly tailored T_r evolution, shown in Fig. 2 for the plastic-ablator capsule described above. Due to the complex hohlraum processes at work, the laser power evolution is even more complicated (see Fig. 2). The requirement of highly-tailored laser pulses, with differing details for different target, place heavy pulse-shaping demands on NIF.

Emphasis on target design is shifting towards ignition-sensitivity studies of the different target designs. These robustness studies can be divided into four categories:

- capsule studies: capsule surface imperfections, volume imperfections, material imperfections, and coupling of these imperfections with asymmetry of imperfect radiation drive

hohlraum studies: variations in the hohlraum length, diameter, laser-entrance-
hole diameter and lining material, gas-fill density, and hohlraum wall material
- laser studies: determvariations in the laser pointing, laser energy and power,
laser pulse shape, and beam balance
- modeling: sensitivity of to the use of particular physics packages and
algorithms, and assumptions and neglected physics in the modeling of the
target performance.

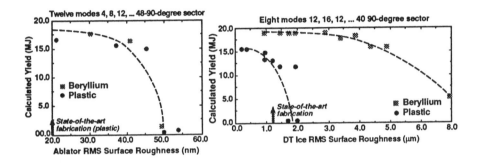

FIGURE 3. The yield from 2-dimensional LASNEX calculations is plotted versus ablator
outer roughness (left) . The yield versus interior-surface roughness of the D-T ice is also
plotted (right).

A detailed review of the extensive studies under way is beyond the scope of this
paper. As an example, a sensitivity study of capsule yield versus ablator-surface
roughness is discussed. This roughness is the seed for Raleigh-Taylor instability
which could induce unacceptable levels of mix. Both the plastic and beryllium
ablators have been studied. The technique used to simulate the instability is "nonlinear multimode modeling", where direct 2-D LASNEX simulation of the effect of
a realistic surface roughness on a capsule surface is calculated with a minimum of
modeling approximations. Results of the study are shown in Fig. 3. The results
indicate that a 50 nm RMS outer surface roughness is tolerable to achieve ignition.
Current Nova capsules have a surface roughness as low as 20 nm, so targets
currently meet this specification. The cliff in target performance for the inner ice
surface is calculated to be 1.5 µm for the CH-ablator target. The advantage of the
Be target on this score is noteworthy.

An important consideration for NIF targets is the feasibility of fabrication. In particular, most ignition-capsule designs rely on a cryogenic layer of DT which must be very smooth, as discussed above. Fortunately, the melting of the ice due to heating from the radioactive β decay of T and subsequent refreezing provides a path for achievement of the required smoothness[5]. In a toroidal test cell, a D-T layer of thickness 110 μm has been deposited with a surface roughness (sum in quadrature from all modes 1--110) of only 1.2 $\pm$ 0.3 μm[6]. Given the requirements discussed above, this represents a very encouraging result.

3. EXPERIMENTAL TARGET-PHYSICS CAMPAIGNS

Present work at Nova has concentrated in completing the "Nova technical contract" originally laid out in 1990 by the review panel from the National Academy of Sciences chartered by the US DOE to review the ICF program. Summary of the status at present status is provided. There are three main issues being presently addressed:

Drive symmetry: Given the relatively high convergence ratio needed for ignition of NIF capsules, a high degree of illumination symmetry is needed. Current estimates indicate the time-averaged symmetry needs to be $\leq$ 1%. Instantaneous asymmetries can be higher than that value. Excellent symmetry control and modeling (with LASNEX) has been demonstrated in both vacuum and lined Nova hohlraums using various diagnostic techniques[7]. However, recent symmetry experiments in gas-filled hohlraums have generally shown anomalously pole-hot capsule illumination. These asymmetries are consistent with a beam-steering effect early in time of about 0.2 mm (compared to a 0.5--0.7 mm beam footprint) as measured by both implosion and re-emission ball techniques[8,7]. It appears that this steering is the result of the deflection of hot spots undergoing mild filamentation[9] by strong plasma flows transverse to the propagating beam[10]. Future experiments will evaluate whether beam smoothing techniques such as finite bandwidth, by virtue of suppressing filamentation, can control this steering effect.

Laser-plasma instabilities: Potentially high levels of parametric laser-plasma instabilities [9] are possible because of the relatively long scale lengths predicted within NIF hohlraums. Laser scattering by unstable hohlraum plasmas are a potential problem for hohlraum energetics (in the case of back scatter) and symmetry (in the case of oblique scatter).

An extensive study of Stimulated Brillouin scatter (SBS), where the laser decays into an ion acoustic wave and scatteed light, has been done in both open targets[11] and hohlraums[12,13] designed to approximate plasma conditions with the longest scale lengths in gas-filled NIF hohlraums. It is found that properly chosen mixtures of ion species (such as He/H for NIF hohlraums) can suppress SBS.

Hydrodynamic stability: The implosion of the ICF capsule is susceptible to the Rayleigh-Taylor (RT) instability during the ablation process and also during the deceleration. Nevertheless, high convergence implosions have been achieved on Nova and successfully modeled[14]. For studying single-mode RT in convergent geometry with good access for time-dependent diagnosis, implosion of foam-filled thin metal cylinders is now used[15]. Implosion of capsules with laser-ablated pits, providing a broad mode-number spectrum, are also used to study convergent RT.

The authors gratefully acknowledge the hard work by the members of the LLNL and LANL ICF programs. This work is supported by the US DOE.

5. REFERENCES

[1] J. H. Nuckolls, L. Wood, A. R. Thiessen, and G. B. Zimmerman, Nature **239**, 139 (1972).

[2] E. M. Campbell, Rev. of Sci. Instrum. **57**, 2101 (1986).

[3] G. Zimmerman and W. Kruer, Comments of Plasma Physics and Controlled Fusion **2,** 85 (1975).

[4] S. W. Haan *et al.,* Phys. of Plasmas 2, 2480 (1995).

[5] J. K. Hoffer and L. R. Foreman, J. of Vac. Sci. Tech., <u>A7</u>, 1161 (1989).

[6] J. K. Hoffer, private communication.

[7] A. A. Hauer, Phys. Plasmas **2**, 2488 (1995).

[8] N. D. Delamater, private communication.

[9] W. L. Kruer, "The Physics of Laser-Plasma Interactions", Addison Wesley, Reading, MA, 1988.

[10] H. A. Rose, E. A. Williams, D. Hinckel, private communications.

[11] B. J. MacGowan, *et al.*, Proc. 15th Int. Conf. Contr. Nucl. Fus. Res., IAEA, Vienna, 1994.

[12] J. C. Fernandez, *et al.*, Phys. Rev. Lett. (1995) submitted.

[13] L. V. Powers, *et al.*, Phys. Rev. Lett. **74,** 2957 (1995).

[14] M. Cable, *et al.*, Phys. Rev. Lett. **73,** 2316 (1994).

[15] W. W. Hsing, Phys. Rev. Lett. (1995) submitted.

**ROBERT L. McCRORY
1995**

Teller Award Acceptance Speech

Professor Robert L. McCrory
Laboratory for Laser Energetics
University of Rochester

It is indeed an honor to receive an award named for such an accomplished and famous physicist who is present with us today, Dr. Edward Teller. In thinking over what to say on this occasion, I noted that the Teller Award was given for pioneering research in controlled fusion, in controlling fusion for the benefit of mankind. I think everyone in this audience certainly would agree that this lofty goal is truly one of the unconquered, grand challenges in applied physics.

I have devoted most of my career to the pursuit of controlled fusion via the inertial confinement fusion approach with lasers. I take this opportunity to note that the roots of the inertial confinement fusion program span more than 40 years. In 1952 the first hydrogen bomb explosion marked the beginning of the thermonuclear age. In fact, that event in the Pacific was really the first "public" announcement that the fusion reaction can be made to occur on earth. The laser was invented in 1960, but if you talk to Dr. John Nuckolls and a number of others active in the thermonuclear weapons program in the United States at the time, the laser provided a solution to a problem they were already working on--the intense energy source, a "match," that was needed to ignite a laboratory-scale microfusion reaction. The earliest concepts for laboratory inertial confinement fusion appear in the literature in papers by Kidder, Dawson, Basov, and Khrokhin. The first neutrons generated by a laser plasma were reported by Basov et al. in 1968. The early 1970s. saw the rapid rise of laboratory programs in inertial confinement fusion at LLNL, LANL, the University of Rochester, SNL, NRL, Osaka, Limeil, Garching, and Rutherford as well as several laboratories in the Former Soviet Union.

In 1972 the fact that we used inhomogeneous or "shell" targets was classified in the United States. Many of you now working in this field may not remember the time when we couldn't talk about structured fusion capsules. If you reexamine the 1972 Nature article by John Nuckolls et al., you will find that the target was a homogeneous spherical fluid shell of DT imploded with a highly shaped pulse. Such a pulse would be very difficult to achieve in the laboratory.

The late 1970s saw a number of multiterawatt lasers marshaled to explore inertial fusion: the 2-TW Argus glass laser at LLNL, followed by the 10-TW Shiva facility; the 10-TW Helious CO_2 facility at LANL; and the 10-TW OMEGA facility, a glass laser at the Laboratory for Laser Energetics, University of Rochester. Early research with these facilities revealed that infrared light and long wavelength light is unsuitable for controlled fusion. At my Laboratory, in 1979 efficient frequency-tripling was invented and demonstrated. That "breakthrough" is now the basis for optimism and incorporated in the design of almost all large glass fusion lasers. The early 1980s saw a number of ultraviolet laser-matter interaction experiments that proved the efficacy of shorter wavelengths. The first ultraviolet experiments were demonstrated at Ecole Polytechnique using frequency-quadrupled light. Larger lasers, such as OMEGA and Nova were converted to the ultraviolet in pursuit of the goals of inertial fusion. In 1986 indirect drive experiments at LLNL using the frequency-tripled Nova system produced nearly 100 times liquid density in compressed thermonuclear fuel.

The experiments conducted world-wide showed the need to have smoother laser beams. Professor Kato at ILE was the first to propose use of a random phase plate to obtain smoother laser beams. The Naval Research Laboratory introduced induced spatial incoherence. In 1988 smoothing by spectral dispersion was invented and demonstrated at LLE. This technique combines the ideas of band width introduced by NRL and beam smoothing using random phase plates introduced by Kato. Using these techniques, LLE was the first to demonstrate compressions near 200 times liquid density in compressed DT with direct drive. At 'the same time, the Halite-Centurion program, a program in the United States using underground explosives laid to rest :many fundamental concerns concerning the feasibility of igniting small masses of DT.

Following the progress of the late 1980s, the 1990s began in the spirit of optimism and openness. In 1994 the United States declassified much of the physics and results of indirect-drive targets world-wide. Plans for larger lasers to reach ignition and gain began. In April 1995, the OMEGA Upgrade, a 45-KJ, ultraviolet, 60-beam, direct-drive facility was completed. In the United States we also saw the beginning of the National Ignition Fusion Project, a billion dollar glass laser that will be used for both direct and indirect drive to demonstrate ignition and gain in the laboratory.

The brief sketch I have just provided shows that the energy scaling of solid-state lasers is based on a solid 25-year record of successful system design, construction, and operation, as illustrated in Fig. 1. A summary of the success of short-wave-length lasers to provide high absorption and low suprathermal electron preheat is shown in Fig. 2 where data from a variety of laboratories world-wide is plotted. The smoothing by spectral dispersion scheme is illustrated in Figs. 3 and 4. Table 1 illustrates the major capsule implosion milestones were achieved on Nova and OMEGA inertial confinement fusion experiments in 1987 and 1988. The table aptly illustrates the success of both indirectly and directly driven capsules.
A disappointment of direct-drive experiments in the late 1980s showed there was a lack of uniformity that could not be addressed by using only 24 beams and inadequate beam smoothing techniques. Figure 5 illustrates that the initial OMEGA experiments suffered from significant pusher fuel mix. This mix significantly degrades the fuel burn and as illustrated in the figure, only about 1/1000 of the ideal yield was realized. Much of my academic published research has been in the area of mix in fusion targets. Figure 6 illustrates the evolution of a devastating Rayleigh-Taylor instability. This instability can be controlled with smoothing techniques and careful pulse shaping to control the isentrope of the imploding capsule. It is both the experimental results and increasing numerical simulation sophistication that has led inertial confinement fusion researchers to have high confidence that with adequate control of the uniformity of drive and fuel isentrope, high performance will be achieved on the National Ignition Facility.

A major purpose of the new OMEGA facility is to experimentally demonstrate irradiation uniformity and near-ignition conditions for directdrive targets. Figure 7 illustrates the configuration of the new OMEGA Upgrade facility. Figure 8 illustrates the target bay of the new facility, and Fig. 9 shows the laser bay. Figure 10 is an artist's conceptual design of the National Ignition Facility that will follow both Nova and OMEGA to demonstrate gain.

In summary then, the pursuit of inertial fusion with glass lasers that has occupied most of my career for over 20 years looks exceptionally bright. I am proud of the contributions the Laboratory for Laser Energetics has made toward demonstrating the scientific feasibility of inertial confinement fusion to date. Major progress on indirect drive on Nova and initial experiments with direct drive on OMEGA will support the final stage in the quest to demonstrate controlled fusion. These experiments will culminate with research planned for the National Ignition Facility. My career in fusion research has been rewarding and I believe Dr. Teller should take great pride as one of the original fusion pioneers that the quest for laboratory-scaled fusion is near at hand.

Major capsule-implosion milestones were achieved on Nova and OMEGA ICF experiments in 1987–1988

UR
LLE

Nova radiatively driven capsules	OMEGA directly driven capsules
$n\tau \sim 2 \times 10^{14}$ cm^{-3} s	$n\tau \sim 2 \times 10^{14}$ cm^{-3} s
$n = 5 \times 10^{24}$ (100 XLD)	$n = 0.5 - 1.0 \times 10^{25}$ (100 – 200 XLD)
$T_i \sim 1.7$ keV	$T_i \leq 1.0$ keV
neutron yield ~2–10^{10}	neutron yield ~10^8
Nova input: 20 kJ; 0.35 μm, 1 ns	OMEGA input: 1.3 kJ; 0.35 μm; 0.6 ns
• demonstrated near 1-D performance at capsule convergence ratio of 25–30	• demonstrated first direct measurement of highly compressed DT fuel
• demonstrated drive symmetry control of ±2.5%	• demonstrated high-density compression with solid-fuel-layer (cryo) targets

E4945

Table I

1326

The energy scaling of solid-state lasers is based on a 25-year track record of successful system design, construction, and operation

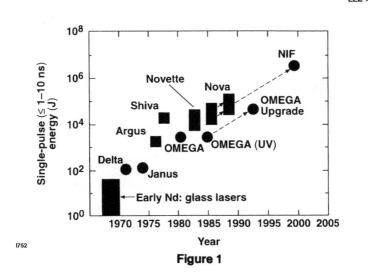

Figure 1

Short-wavelength lasers provide high absorption and low superthermal electron preheat

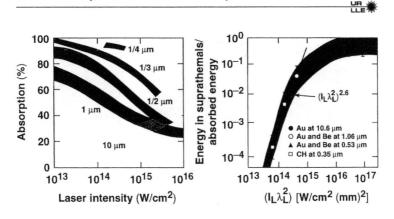

*Data from LLE, LLNL, Osaka, LANL, NRL, and Ecole Polytechnique.

Figure 2

SSD (smoothing by spectral dispersion) provides uniform target illumination for frequency-converted laser systems

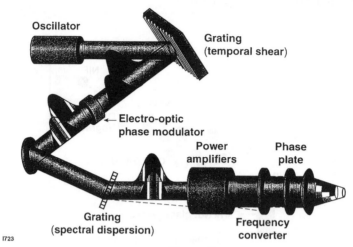

I723

Figure 3

Broadband phase conversion (SSD) provides ultra-uniform irraidation

- Equivalent-target-plane photographs
- Phase conversion with distributed phase plates
- Broadband spectrum = 2.1±0.15Å

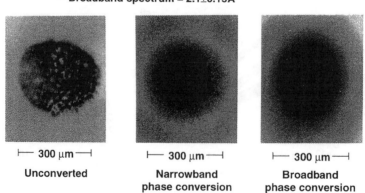

⊢— 300 μm —⊣	⊢— 300 μm —⊣	⊢— 300 μm —⊣
Unconverted	Narrowband phase conversion	Broadband phase conversion

I722

Figure 4

OMEGA 200 XLD cryogenic target experiments indicate significant pusher-fuel mix

UR
LLE

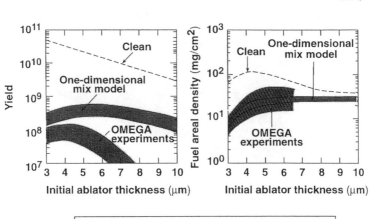

Glass-DT targets have an interfacial instability not present in high-gain designs.

I726

Figure 5

At late stage, few bubbles are controlling the front evolution and punching the shell

UR
LLE

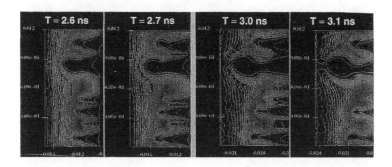

TC3703

Figure 6

The OMEGA laser is designed to achieve high uniformity with flexible pulse-shaping capability

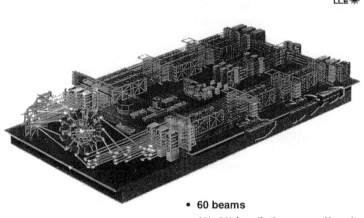

- 60 beams
- 1%–2% irradiation nonuniformity
- Flexible pulse shaping
- Short shot cycle (1 h)

TC2998a

Figure 7

The target bay is in final stages of completion

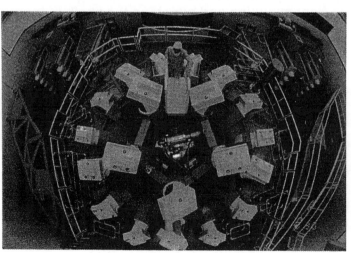

I940

Figure 8

The laser is in final activation for Key Decision IV

I941

Figure 9

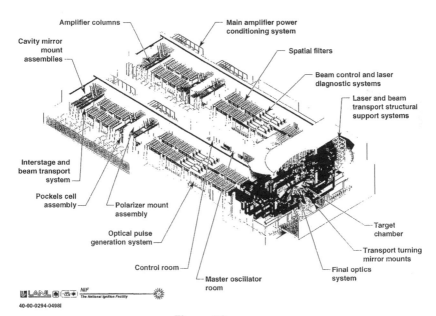

Amplifier columns

Cavity mirror
mount
assemblies

Main amplifier power
conditioning system

Spatial filters

Beam control and laser
diagnostic systems

Laser and beam
transport structural
support systems

Interstage and
beam transport
system

Pockels cell
assembly

Polarizer mount
assembly

Optical pulse
generation system

Control room

Master oscillator
room

Target
chamber

Transport turning
mirror mounts

Final optics
system

NIF
The National Ignition Facility

40-00-0294-0498I

Figure 10

GEORGE H. MILEY
1995

1995 Edward Teller Lecture
Patience and Optimism

George H. Miley

Fusion Studies Laboratory • University of Illinois
100 NEL, 103 South Goodwin Avenue
Urbana, IL 61801-2984 USA

Abstract. Remarks made in the author's acceptance lecture for the 1995 Edward Teller Medal are presented and expanded. Topics covered include research on nuclear-pumped lasers, the first direct e-beam-pumped laser, direct energy conversion and advanced fuel fusion, plus recent work on inertial electrostatic confinement. "Patience" and "optimism" are viewed as essential elements needed by scientists following the "zig-zag" path to fusion energy production.

INTRODUCTION

It is a great privilege to receive an award associated with Dr. Edward Teller. His many accomplishments, insight, and leadership for over half a century have been an inspiration for scientists worldwide. His personal involvement in the Awards Ceremony was especially appreciated.

I have titled this presentation "Patience and Optimism," feeling that these characteristics have been important not only to my own career, but that they must also be the cornerstone of mankind's development of fusion energy. That is true despite the fact that optimism, if overdone, begets impatience, which, in turn, can be destructive in scientific endeavors.

Before continuing, I wish to acknowledge the strong support that I have received throughout my scientific career from my family and also from the many student research assistants who have completed M.S. and Ph.D. theses with me since I arrived at the University of Illinois (UI) in 1961. My research has greatly benefited from the opportunity to teach and collaborate with these talented students.

A focal point of my research over the years has been to steer fusion development towards a maximum environmental compatibility. This goal, in my view, requires the timely development of advanced-fuel (e.g. D-^{3}He) fusion, along with efficient advanced (including "direct") energy conversion. These features, combined with the use of low-activation materials, will provide limited radioactivity, a minimum radioactive inventory, and maximum efficiency. Such characteristics will be essential for an attractive power source by the time fusion is ready to enter the commercial energy market.

NUCLEAR-PUMPED LASER RESEARCH

Nuclear-pumped lasers (NPLs) have been a major research interest for me for many years, and, as pointed out later, NPLs are viewed as a unique high-efficiency inertial confinement fusion (ICF) driver. The concept of using energy released in neutron-induced nuclear reactions in the laser media to "pump" the laser occurred to me in 1963, after reading one of the first papers on electrically pumped lasers. As it turned out, however, the concept had been advanced several months earlier by L. Herwig at the United Aircraft Research Lab.[1] Subsequent calculations led me to believe that a CO_2 NPL could be achieved with a very low neutron flux threshold, readily achievable with the UI's pulsed TRIGA research reactor. Thus, my collaborators and I undertook a series of experimental studies of this system, only to find that it would not work at all, probably due to excessive dissociation of CO_2 in the intense radiation field. As a fall-back position, we developed a radiation-sustained electrical discharge CO_2 laser,[2] which, in its own right, provided an important new direction for CO_2 laser research. Returning to NPLs, we next initiated research on a He-Ne NPL, feeling that the physics for this laser was well-understood, and calculations predicted lasing. That was not the case experimentally, however. This effort did, though, provide an important widening base for understanding radiation-induced plasmas involving noble gases.[3,4] Thus, while much basic information had been achieved, we did not obtain an NPL in this research effort until 1976, when we announced a Ne-N_2 laser[5] and, shortly thereafter, the first visible NPL based on a resonant charge-exchange mechanism in He-Hg.[6]

A succession of new NPLs followed, including the discovery of the family of "impurity"-type lasers, operating on C and N lines in He mixtures. [7] A major new development, reported at an international meeting on NPLs in Paris in 1978,[8] was our measurement of gain in several excimers, in agreement with theory. Pumping powers available were too low for lasing, however, and this limitation later led to the development of excimer-based nuclear-pumped "flashlamps."[9] (Pulsed research reactors characteristically provide a large energy input into the laser medium, but due to the relatively wide pulse, deliver a low peak power compared to pulsed electrical pumping.)

In the 1970s, the main goal of UI NPL research was directed at development of an NPL for use as an ICF driver. This interest was motivated by the realization that use of a neutron-feedback NPL-driven ICF reactor (i.e. neutrons from a target burn would pump the NPL, which in turn "fires" on the next target) would provide a high-efficiency power plant.[10] Two key NPL systems were studied for this use. The first such laser used nuclear pumping of oxygen to generate O_2 ($^1\Delta$). This was demonstrated experimentally (Fig. 1), and a laser scheme using a combination O_2 ($^1\Delta$) generator, flashlamp decomposition of O_3 (for driving an I_2 laser) was examined computationally.[11] A key feature of this NPL is the long (>msec) storage time afforded through the O_2 ($^1\Delta$), satisfying the need for a large energy storage time between target burns. Another approach to an ICF driver was reported in 1992,[12] where a nuclear-pumped XeBr flashlamp was employed to photolytically pump iodine (Fig. 2). This laser would not provide long energy storage, unlike O_2 ($^1\Delta$), but could be used in a hybrid fission-reactor-driven NPL-ICF power plant.[13] The hybrid approach offers considerable flexibility and also provides good energy efficiency, but it does, however, introduce the complication of increased radioactivity from the fission component.

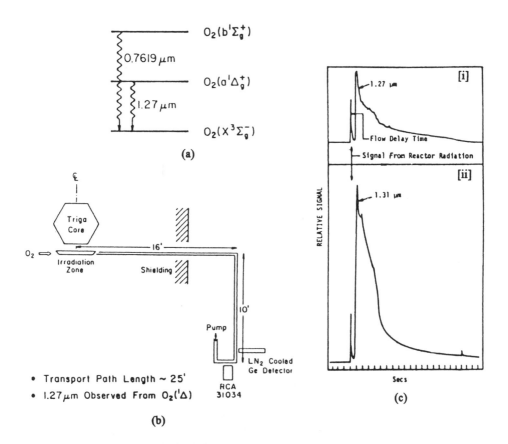

FIGURE 1.   First experimental observation of nuclear pumping of $O_2(^1\Delta)$. (a) simplified energy level diagram for oxygen; (b) schematic of experimental apparatus; (c) the radiation signal observed from [i] the O_2-Ar mixture 0.8 sec after TRIGA pulse, and [ii] from the O_2-Ar mixture after mixing with iodine.

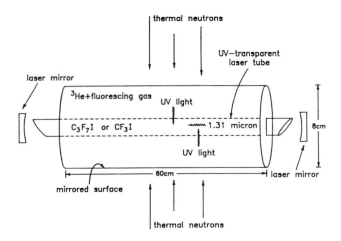

FIGURE 2.   Iodine laser driven by XeBr nuclear-pumped flashlamp.

In the 1970s, the National Aeronautics and Space Administration (NASA) became interested in the use of NPLs for space communication. This led NASA scientists to undertake an extensive study of a variety of noble gas-based NPLs operating in the near-infrared (IR) range,[14] culminating in the record achievement of a 1-kW "box"-type NPL by a former UI researcher (R. DeYoung).

Spurred on by the various successes in NPL research, several major programs for development of NPLs for space power beaming emerged in the 1990s. Following a presentation by the author at the *16th Winter Colloquium on Quantum Electronics* in 1986,[15] workers from Lawrence Livermore National Laboratory (LLNL) initiated a scale-up study, designated "Centaurus," using INEL-Idaho facilities.[16] INEL staff were heavily involved, and eventually the Sandia National Laboratories (SNL) "FALCON" project was created, with the goal of demonstrating a multi-kW Ar-Xe NPL.[17,18]

The end of the "Cold War" led to an important series of international conferences on NPLs in Russia: NPL '92 and NPL '94.[19,20] These meetings, held in Obninsk and Arzamas-16, respectively, revealed that Russia had developed a major NPL program, including a number of unique pulsed reactors to drive the NPLs. These meetings also revealed that Soviet scientists had closely followed the UI NPL research and had duplicated a number of our experiments.

In 1992-3, a series of experiments at UI led to two significant results: a new, low-threshold visible NPL employing He-Ne-H_2, and the first B^{10}-pumped Ar-Xe NPL.[21,22] Recent UI studies of NPLs, reported at the present (*12th International Conference on Laser Interaction and Related Plasma Phenomena*) conference,[23,24] have concentrated on control of thermal focusing effects to improve NPL beam quality and on radiation effects on optical materials.

In conclusion, others in the NPL area also deserve note for their many research contributions. In the US, these include my early UI collaborator, J.T. Verdeyen; D. McArthur and colleagues at SNL; K. Thom, F. Hohl and R. DeYoung at NASA; V. George and colleagues at LLNL; K. Watts and staff at INEL; R.T. Schneider and others at the University of Florida; and H.H. Helmick at Los Alamos National Laboratory, to name a few. In Russia, A. Zradnikov and his staff at Obninsk, A. Voinov and his staff at Arzamas-16, and E. Magda and his colleagues at Chelyabinsk-70 have obtained outstanding results.

FIRST DIRECT E-BEAM PUMPED LASER

A diversion from NPL research occurred in 1969-70, with very exciting results. I had been invited to visit Cornell University for one year to teach a course on fusion technology and to study the potential for GeV ion acceleration, using a collective wave technique and their new relativistic electron beam facility. The ion acceleration project was not showing progress, so I decided to undertake electron-beam pumping of various gas laser systems. This highly successful project produced the first "direct" e-beam-pumped laser (Fig. 3),[25] although a traveling-wave-type e-beam laser had been achieved earlier at the Naval Research Laboratory (NRL).

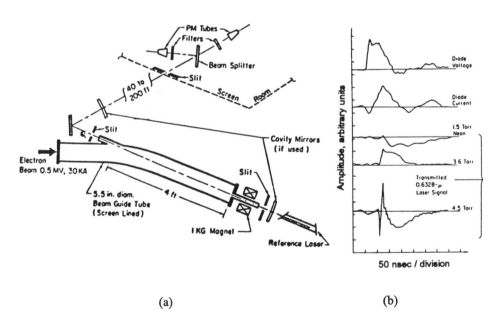

(a)   (b)

FIGURE 3.   Laser excitation with a relativistic electron beam; (a) the experimental set-up; (b) current-voltage traces and 6238-Å gain measurements for several neon pressures. (From Ref. 25)

DIRECT ENERGY CONVERSION and
ADVANCED FUEL FUSION

The book, <u>Direct Conversion of Radiation Energy</u>,[26] written in 1970 under Atomic Energy Commission sponsorship, in a sense provided a "bridge" between NPLs and fusion research concepts (Fig. 4). This book traced the division of radiation-deposited energy between ionization and excitation of the absorbing media (Fig. 5). The potential of lasing (e.g. the NPL) was discussed as a means of directly extracting excitation energy, while various concepts for "ionization-electric" cells were presented, including the gamma electric cell (direct conversion of Compton electron energy to electricity), the fission electric cell (direct conversion of fission-fragment energy to electricity), and the "Post-type" electrostatic direct energy convertor for mirror-type fusion reactors. Several important near-term applications developed from this work. The gamma electric cell concept (Fig. 6) was subsequently exploited by a group at LANL to produce ultrahigh voltages. Also, a variation of the fission electric cell emerged as a series of commercial neutron and gamma detectors (Semirad[TMa] detectors) for use in high radiation fields.

DIRECT
CONVERSION
OF NUCLEAR
RADIATION ENERGY

George H. Miley
University of Illinois

This book presents for the first time a comprehensive study of methods for converting nuclear radiation directly without resorting to a heat cycle.

The concepts discussed primarily involve direct collection of charged particles released by radioisotopes and by nuclear and thermonuclear reactors.

© American Nuclear Society, 1970

FIGURE 4.   Cover and overleaf description from *Direct Conversion of Nuclear Radiation Energy*, 1970--the first comprehensive overview of this topic.

---

[a] Reuter Stokes Company, Inc. trademark

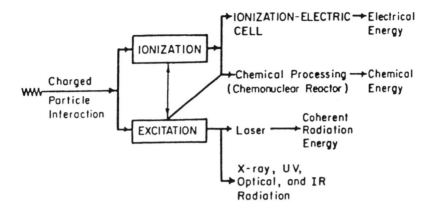

FIGURE 5. The relationship of interaction-energy cells developed in the 1970 book on direct conversion of nuclear radiation energy. (Ref. 26)

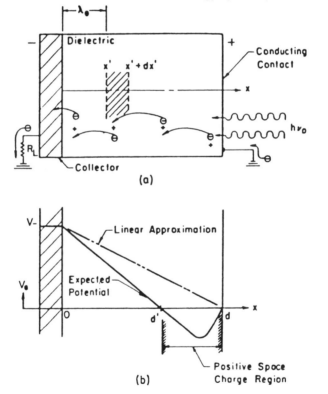

FIGURE 6. The Gamma Electric cell; (a) electron transport through the dielectric; (b) the resulting potential. The maximum potential obtainable is limited only by the leakage current, due to a finite resistivity of the dielectric employed. (Ref. 26)

Interest in direct conversion, combined with fusion, led to the 1976 book, Fusion Energy Conversion,[27] sponsored by the Energy Research and Development Agency, the predecessor of the U.S. Department of Energy. (Fig. 7) In preparing this book, I was forced to face a new and very important issue tied to efficient fusion energy conversion--the need to develop use of advanced fuels, such as D-^{3}He, in order to obtain a large fraction of the fusion energy in the form of charged fusion products. This in turn led to a further examination of issues associated with advanced fuels, e.g. sources of ^{3}He, radiation losses at high temperature, methods to handle large first-wall heat loads, and most importantly, the search for a plasma confinement approach compatible with these fuels. Advantages of advanced fuel fusion--increased efficiency, reduction of neutron damage to the first wall and structural components, and reduction of tritium and induced radioactivity--must be considered as part of the trade-off vs. use of more easily burned D-T. This 1976 book, then, provided a first comprehensive look at all of these issues.

In addition to fuel cycles, a variety of energy conversion methods were presented, including a brief section devoted to direct conversion of ICF micro-explosion energy by expansion against a magnetic field (Fig. 8). Another topic covered in this book, non-electricity-producing uses of fusion reactors, has not received much attention to date, but appears to be growing in importance in view of the need to provide the broad base of energy alternatives required to meet future applications.

While considered too visionary by some when it was first published, Fusion Energy Conversion identified a field of interest that has continued to grow and has gradually commanded more attention by those planning for the future applications of fusion energy. Subsequently, the realization that lunar mining offers a plentiful source of ^{3}He helped to focus new attention on the potential of D-^{3}He fusion.[28] However, breeding ^{3}He, e.g. using a D-D "breeder" with D-^{3}He satellite reactors, also potentially offers a very attractive energy system (Fig. 9).[29,30,31] If a good method to burn p-^{11}B is found, the advanced-fuel resource issue would be removed, but utilizing this fuel will require a major breakthrough in confinement technique.

FIGURE 7. *Fusion Energy Conversion*, 1976, represented the first comprehensive discussion of the need to develop advanced fuel fusion.

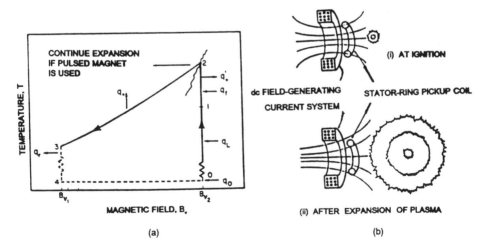

FIGURE 8. Examples of electromagnetic coupling schemes; (a) the ideal thermodynamic cycle for a fusion-power system, and (b) the operation of ac-MHD conversion with ICF micro-explosions. (Ref. 27)

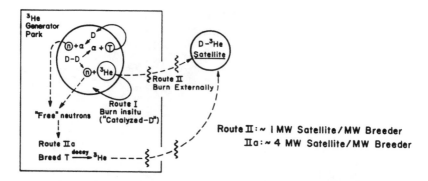

FIGURE 9.   The D-^{3}He Satellite concept--three options for use of ^{3}He extracted from a
semi-catalyzed D reactor. Option (Route) I recirculates ^{3}He, giving a fully
catalyzed D reactor. Route II delivers ^{3}He to a separate D-^{3}He reactor.
Route IIa breeds tritium, which then decays to ^{3}He for use in the external
reactor. Route II has the advantage of avoiding a large tritium stockpile, but
results in a lower satellite/breeder electrical energy ratio. Still, Route II
appears very attractive for future combinations of synthetic fuel, which
would be produced in the breeder reactor, and localized electrical production from the D-^{3}He satellite reactors.

Following a series of D-^{3}He magnetic confinement studies (largely focused
on the field-reversed configuration [FRC] as well-suited to D-^{3}He fuel, due to the
combination of closed inner field lines for good confinement with open outer field
lines for coupling to a direct converter),[32,33] concepts for using advanced fuels
with ICF were developed. There are two primary challenges involved: the larger
energy input required for ignition and the reduced burn-up fraction, due to the lower
fusion cross section, relative to D-T fusion (Fig. 10). The first concept to be
explored was to shock heat a small D-T "core," which would burn propagate into a
surrounding fuel layer (see Fig. 11). This approach reduces the driver input-energy
requirement, but still requires a somewhat larger ρR target than pure D-T. Several
different outer-region fuel compositions are possible: pure D, D-^{3}He, etc. A special
goal in an initial study was to use a D outer region, designed such that the D-D
tritium produced and not burned (generally that implanted in the outer shell) would
just equal the tritium burned in the D-T "spark" core (see Fig. 12).[34] Such a
tritium "self-contained" target was named AFLINT. To investigate the advantages
of the AFLINT target, a reactor study, sponsored by the Electric Power Research
Institute (EPRI), was undertaken. As illustrated in Fig. 13, LOTRIT used a unique
twin lead "fall" concept.[35] Lead coolant was used instead of lithium, since external
tritium breeding was not required; the thin inner fall was vaporized by each target
micro-explosion, the debris and vapor being condensed out on the thick outer fall.
This configuration allowed a more compact reaction chamber than the earlier D-T
Hiball reactor design.[35-37] As implied by the name, LOTRIT also offered a smaller
tritium inventory than prior D-T designs by several orders of magnitude.

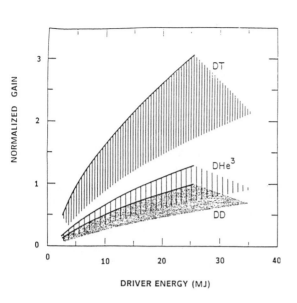

FIGURE 10. Comparison of estimated
gain curves for advanced
fuel vs. D-T ICF targets.

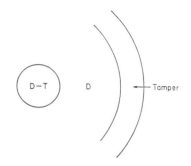

FIGURE 11. The AFLINT target.

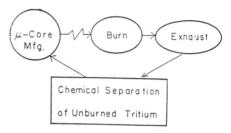

FIGURE 12. Flow diagram for tritium
produced by D-D reac-
tions in AFLINT.

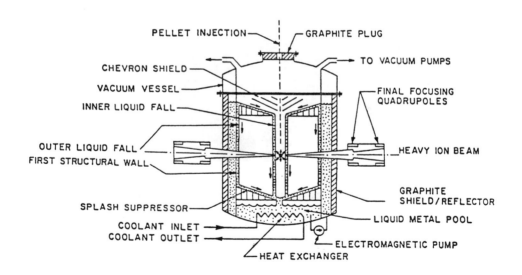

FIGURE 13. LOTRIT reactor chamber vertical cross section.

A later version of LOTRIT incorporated a novel two-phase liquid metal magnetohydrodynamic (MHD) energy converter to increase the overall plant efficiency.[36] Subsequent advanced fuel studies examined variations of the AFLINT concept, largely designed to include ^{3}He.[34,37] Also, collaborative studies on the alternative approach of volume ignition of D-^{3}He and p-^{11}B were reported.[38]

NEW DIRECTIONS: IECs AND DPFs

Recent work has turned to a "rebirth" of research on inertial electrostatic confinement (IEC).[39,40,41] As illustrated in Fig. 14, ions are extracted from a plasma discharge formed between a spherical grid and the vacuum vessel wall. High-voltage (~50-80 kV) acceleration of the ions towards the center of the spherical vessel results in beam-beam fusion reactions in a central high-density plasma core. Electrons, interacting with the central space charge, create a complex potential well structure (defined by Farnsworth as a "Poissor"[42,43]) that traps high-energy ions, further increasing the fusion rate. At present, we have developed small IEC devices to serve as portable neutron sources (~10^6-10^7 D-D n/sec) for activation analysis applications.[40] Higher-intensity versions are planned for advanced neutron and proton source applications.[41] Also, design studies for space power applications (Fig. 15) and for H$_2$ production have appeared.[44,45]

Additional work, aimed at space power and propulsion, has involved dense plasma focus (DPF) experiments.[46,47] The main objective has been the development of a gas-injected version of the DPF, with D-^{3}He operation as an ultimate goal. Experiments have been carried out on a 50-kJ DPF, while a larger 250-kJ unit is under construction.[48]

FIGURE 14. Artist's sketch of cut-away IEC vacuum chamber, showing the spherical vessel, ceramic insulator, and inner wire cathode grid.

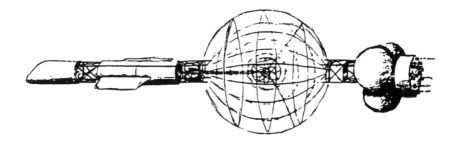

FIGURE 15. Artist's sketch of a spacecraft for a Mars mission, with a D-^{3}He IEC direct energy converter to power electric arcjet thrusters. The IEC and its surrounding grids are located in the center of the sketch, with the crew cabin on the left and electric thrusters on the right.

CONCLUSION

While the research described here has varied greatly in detail, the common link has been the focus on development of an attractive long-term energy source. Fusion energy is plentiful in space, and clearly it will eventually be a major resource for mankind's use on earth. However, both public supporters and researchers must realize the importance of a persistent, deliberate attack for its successful development. Progress by small but cumulative steps has been very successful to date. If maintained, this route should eventually lead us to a most attractive, versatile, and inexhaustible energy resource. Unfortunately, that route has not proven to be a straight path--it has taken many twists and turns. For example, severe cuts in the U.S. DOE budget for fusion research appear likely this year. The politicians who control funding change from year to year, as do national financial priorities. Already, fusion represents the longest continuously funded government research program worldwide, but the viability of such a long-term program is always vulnerable to the typically short-term views of politicians. Yet the goal is so important to future generations that it must be pursued vigorously. "Patience" and "optimism" are essential, if we are to find our way along this "zig-zag" path.

REFERENCES

1. L. Herwig, *C-110053-5*, United Aircraft Laboratories, East Hartford, CT (1964); *Trans. Am. Nucl. Soc., 7*, 131 (1964).

2. T. Ganley, J. Verdeyen, and G.H. Miley, *Appl. Phys. Lett., 43*, 319-327 (1971).

3. J. Guyot, G.H. Miley, and J. Verdeyen, *J. Appl. Phys., 43*, 5379-5391 (1971).

4.   J. Guyot, G.H. Miley, and J. Verdeyen, *Nucl. Sci. and Eng.*, 48, 373-86 (1972).

5.   R.J. DeYoung, W.E. Wells, G.H. Miley, and J.T. Verdeyen, *Appl. Phys. Lett.*, 28, 519-521 (1976).

6.   A. Akerman, G.H. Miley, and D. McArthur, *Appl. Phys. Lett.*, 30, 409-412 (1977).

7.   M.A. Prelas, M.A. Akerman, F.P. Boody, and G.H. Miley, *Appl. Phys. Lett.*, 31, 428-430 (1977).

8.   G.H. Miley, *1st Intern. Symp. on Fission Induced Plasma and Nucl. Pumped Lasers*, Orsay, France, 87-102 (1978); also see G.H. Miley, S.J.S. Negalingam, F.P. Boody, and M.A. Prelas, *Intern. Conf. on Lasers '78*, Orlando, FL, 5-13 (1978).

9.   M.A. Prelas, F.P. Boody, G.H. Miley, and J.F. Kunze, *Laser and Particle Beams*, 6, 1, 25-62 (1988).

10.   G.H. Miley, *Atomkernenergie/Kerntechnik*, 45, 14-18 (1984); G.H. Miley, *XII European Conf. on Laser Interaction with Matter*, Moscow, USSR, 3 pages (December 1978); G.H. Miley, *Energy Storage, Compression and Switching*, Venice, Italy (December 1978); G.H. Miley, E. Greenspan, and J. Gilligan, *Atomkernenergie/Kerntechnik*, 36, (1981); D.E. Beller and G.H. Miley, *Fusion Tech.*, 15, 772-777 (1988).

11.   H.E. Elsayed-Ali and G.H. Miley, *Intern. Conf. on Lasers '83*, San Francisco, CA (December 1983); also see M.S. Zediker, T.R. Dooling, and G.H. Miley, *Intern. Conf. on Lasers '81*, 492-498 (1981); M.S. Zediker, T.R. Dooling, E. Greenspan, and G.H. Miley, *9th Symp. on Engr. Prob. of Fusion Research*, Chicago, IL, II, 1560-1561 (1981); H.E. Elsayed-Ali and G.H. Miley, *J. Appl. Phys.* 60, 1189-1205 (1986); H.E. Elsayed-Ali and G.H. Miley, *Plasma Chemistry and Plasma Processing*, 6, 259-280 (1986).

12.   W.H. Williams and G.H. Miley, *Appl. Phys. Lett.*, 62, 15, (12 April 1993). Also see W.H. Williams and G.H. Miley, *Transactions-Specialist Conf. on Physics of Nuclear Induced Plasmas and Problems of Nuclear Pumped Lasers*, IPPE, Obninsk, Russia, (May 1992); W.H. Williams and G.H. Miley, *Laser and Particle Beams*, 11, 3, 567-573 (1993).

13.   M.A. Prelas and F.P. Boody, *Laser Interaction and Related Plasma Phenomena*, 9, eds. H. Hora and G.H. Miley, Plenum Press, New York, 197-210 (1991); M.A. Prelas and F.P. Boody, *Laser Interaction and Related Plasma Phenomena*, 10, eds. G.H. Miley and H. Hora, Plenum Press, New York, 67-78 (1992); L.-T.S. Lin, M.A. Prelas, Z. He, J.T. Bahns, W.C.

Stwalley, G.H. Miley, E.G. Batyrbekov, Y.R. Shaban, and M. Petra, *Laser and Particle Beams*, 13, 1, 95-109.

14. R. DeYoung, *Appl. Phys. Lett.*, 38, 297 (1981); K. Thom and R.T. Schneider, *AIAA Journal*, 10, 400 (1972).

15. G.H. Miley, *16th Winter Colloquium on Quantum Electronics*, Snowbird, UT (January 1986).

16. K. Watts, personal communication, INEL, 1995. All Centaurus reports, classified SRD, are on file with OSTI, ORNL, Oak Ridge, TN.

17. G.N. Hays, D.A. McArthur, D.A. Neal and J.K. Rice, *Laser Interaction and Related Plasma Phenomena*, 7, eds. H. Hora and G.H. Miley, Plenum Press, New York, 133-142 (1986).

18. D.A. McArthur, G.N. Hays, W.J. Alford, D.R. Neal, D.E. Bodette, and J.K. Rice, *Laser Interaction and Related Plasma Phenomena*, 8., eds. H. Hora and G.H. Miley, Plenum Press, New York, 75-86 (1988).

19. Papers for Specialist Conf. on Physics of Nucl. Induced Plasmas and Problems of Nucl. Pumped Lasers, Obninsk, Russia, 26-28 May 1992, *Laser and Particle Beams*, 11, 3-4 (1993).

20. Papers from *Specialist Conf. on Physics of Nucl. Induced Plasmas and Problems of Nucl. Pumped Lasers*, Arzamas-16, Russia, September 24-October 1, 1994. All-Russian Institute of Experimental Physics (VNIIEF). In Press.

21. Y.R. Shaban and G.H. Miley, *Laser and Particle Beams*, 11, 3, 559-566 (1993); G.H. Miley, Y. Shaban, M. Petra, E.G. Batyrbekov, and E. Suzuki, *First Wireless Power Transmission Conf.: The Commercial Potential*, ed. A. Brown, Center for Space Power, College Station, TX, 173-182 (1993).

22. E.G. Batyrbekov, E.D. Poletaev, E. Suzuki, and G.H. Miley, *Laser Interaction and Related Plasma Phenomena*, ed. G.H. Miley, AIP Conf. Proceedings 318, AIP Press, New York, 515-516 (1994).

23. G.H. Miley, M. Petra, E. Suzuki, E. Batyrbekov, E. Poletaev, A. Fedenev, *Abstracts: 12th Intern. Conf. on Laser Interaction and Related Plasma Phenomena*, Institute of Laser Engineering, Osaka, Japan, 191 (1995).

24. G.H. Miley, M. Petra, S.G. DelMedico, O.M. Barnouin, *Abstracts: 12th Intern. Conf. on Laser Interaction and Related Plasma Phenomena*, Institute of Laser Engineering, Osaka, Japan, 96 (1995).

25. G.H. Miley, *IEEE Record of the 11th Intern. Symp. on Electron, Ion and Laser Beam Tech.*, San Francisco Press, San Francisco, 279-290 (1971).

26. G.H. Miley, *Direct Conversion of Nuclear Radiation Energy*, American Nuclear Society and Atomic Energy Commission, Hinsdale, IL (1970).

27. G.H. Miley, *Fusion Energy Conversion*, American Nuclear Society, Hinsdale, IL (1976).

28. L.J. Wittenberg, J.F. Santarius, and G.L. Kulcinski, *Fusion Tech.*, 10, 167-178 (1986).

29. G.H. Miley, *Nucl. Inst. and Methods in Phys. Research*, A271, 197-202 (1988).

30. G.H. Miley, *3rd Miami Intern. Conf. on Alt. Energy Sources*, Miami, FL, (15-18 December 1980).

31. E. Greenspan and G.H. Miley, *Nucl. Tech./Fusion*, 2, 43 (1982).

32. G.H. Miley and J.G. Gilligan, *Energy*, 4, 2, 163-170 (1979); D. Driemeyer, G.H. Miley, and W.C. Condit, *Computational Methods in Nucl. Eng.*, 2, 7-37 (1979); G.H. Miley, *Intern. J. of Energy Systems*, 1, 260-266 (1981).

33. G.H. Miley, W. Kernbichler, and R. Chapman, *Intern. Conf. on Emerging Nucl. Energy Systems*, Karlsruhe, FRG, (July 1989); O. Barnouin, B. Temple, and G.H. Miley, *Fusion Tech.*, 19, 3, Part 2A, 846-851; J.G. Gilligan, G.H. Miley, and D. Driemeyer, *4th Topical Mtg. on the Tech. of Contr. Nucl. Fusion*, King of Prussia, PA, II, 840 (1980).

34. G.H. Miley, C. Choi, and D. Lee, *2nd Intern. Topical Conf. on High Power Electron and Ion Beam Research and Tech.*, I, 243-256 (1977); C.K. Choi, G.H. Miley, D. Lee, and C. Powell, *3rd Topical Mtg. on Tech. of Contr. Nucl. Fusion*, Santa Fe, NM, 405-407 (1978); G.H. Miley, *Proc. of the 6th Intern. Conf. on High-Power Particle Beams* (Beams '86), Kobe, Japan, 309-312 (1986); E. Greenspan, G.H. Miley, and M. Ragheb, FSL-17 (1980).

35. J. Stubbins, B. Adams, M. Ragheb, C. Choi, and G.H. Miley, *Proc. 10th Symp. on Fusion Engineering*, 1901-1905 (1983); G.H. Miley, M. Ragheb, J. Stubbins, and C. Choi, *6th Miami Intern. Conf. on Alt. Energy Sources* (December 1983); G.H. Miley, J. Stubbins, M. Ragheb, C. Choi, G. Magelssen, and R. Martin, *12th Symp. on Fusion Tech.*, Jülich, FRG (September 1982); M. Ragheb, G.H. Miley, and J. Stubbins, *Trans. ANS*, 45, 180 (1983).

36. G.H. Miley, *Trans. ANS*, 52, 262 (1986).

37. G.H. Miley, J.F. Stubbins, M. Ragheb, and C. Choi, *Nucl. Tech./Fusion*, 4, 2/3, 889-894 (1983).

38. R. Khoda-Bakhsh, H. Hora, and G.H. Miley, *Fusion Tech.*, <u>24</u>, 1, 28-36 (1993).

39. G.H. Miley, J. Nadler, T. Hochberg, Y. Gu, and O. Barnouin, *Fusion Tech.*, <u>19</u>, 840-845 (1991); G.H. Miley, J.H. Nadler, and Y.B. Gu, *19th IEEE Intern. Conf. on Plasma Science*, 3P28, 140, Tampa, FL (1-3 June 1992).

40. G.H. Miley, J. Javedani, R. Nebel, J. Nadler, Y. Gu, A. Satsangi, and P. Heck, *Dense Z-pinches*, eds. M. Haines and A. Knight, AIP Conf. Proc. 299, AIP Press, New York, 675-689 (1994).

41. G.H. Miley, A. Satsangi, J. Javedani, Y. Yamamoto, *Seventh Intern. Conf. on Emerging Nucl. Energy Systems* (ICENES '93), ed. Hideshi Yasuda, World Scientific, Singapore, 66-70 (1994); G.H. Miley, J.B. Javedani, Y.B. Gu, A.J. Satsangi, P.F. Heck, I.D. Tzonev, M.J. Williams, R.A. Nebel, D.C. Barnes, L. Turner, J.H. Nadler, and J. Sved, *First Intern. Workshop on Accelerator-Based Neutron Sources for Boron Neutron Capture Therapy*, INEL Conf-940976, 79-88 (1995); I.V. Tzonev, G.H. Miley, R.A. Nebel, *IEEE Conf. Record - Abstracts: 22nd IEEE Intern. Conf. on Plasma Science*, 95CH35796, 258 (1995).

42. P. Farnsworth, U.S. Patent #3,258,402, June 28, 1966; U.S. Patent #3,386,8-83, June 4, 1968.

43. R.L. Hirsch, *J. Appl. Phys.*, <u>38</u>, 11, 4522 (1967).

44. G.H. Miley, R. Burton, J. Javedani, Y. Gu, A. Satsangi, P. Heck, R. Nebel, and N. Schulze, *Vision-21 Interdisciplinary Science and Engineering in the Era of Cyberspace*, NASA Conference Publication 10129, 185-198 (1993); A.J. Satsangi, G.H. Miley, J.B. Javedani, H. Nakashima, and Y. Yamamoto, *Proc. of the 11th Symp. on Space Nucl. Power and Propulsion*, Conf. 940101, AIP Press, 1297-1302 (1994).

45. G.H. Miley, J. Javedani., Y. Yamamoto, Y. Gu, A. Satsangi, P. Heck, R. Nebel. L. Turner, R.W. Bussard, M. Ohnishi, K. Yoshikawa, H. Momota, and Y. Tomita, *Proc. of the First Intern. Conf. on New Energy Systems and Conversions*, ed. T. Ohta and T. Homma, Frontiers Science Series No. 7, Universal Academy Press, Inc. 183-188 (1993).

46. R. Nachtrieb, O. Barnouin, B. Temple, G. Miley, C. Leakeas, C. Choi, and F. Mead, *18th IEEE Intern. Conf. on Plasma Science*, Williamsburg, VA, 155 (1991); S. DelMedico, B. Bromley, J. Javedani, O. Barnouin, G.H. Miley, et al., *Final Report*, Phillips Laboratory, Propulsion Directorate, Air Force Materiel Command, PL-TR-94-3002 (1994).

47. G.H. Miley, R. Nachtrieb, and J. Nadler, *AIAA-91-3617* (1991).

48.   S. DelMedico, J. Javedani, R. Wimmer, D. Neupert, O. Barnouin, R. Burton,
       M. Williams, B. Bromley, J. DeMora, and G.H. Miley, *1994 IEEE International Conference on Plasma Science-Abstracts*, IEEE No. 94CH3465-2, 130
       (1994); M.J. Williams, J.B. Javedani, E.L. Lerner, A. Peratt, and G.H. Miley,
       Bult. APS, 39, 7, 1769 (1994).

GENNADY A. KIRILLOV
1995

Teller Award acceptance speech
Gennady A. Kirillov
Deputy Director
Federal Russian Nuclear Research Institute for Experimental Physics
Arzamas-16, Russia,

Since I am not good at speaking English, I would like to ask Dr.Kochemasov to translate my Russian.

Here I would like to talk shortly about studies on ion-laser physics at Arzamas-16. We have begun these investigations since 1965. The group was under the leadership of Khariton and Kohmir, and I was one of the active scientists. In the beginning, the research was proceeded in collaboration with Basov and Krohkin at Levedev institute. In 1961, it was reported that in terms of molecular dissociation it is possible to achieve inversion population. It was 1965 when a practical ion-laser system was first realized. We then got started on the research with the following experimental set up. First, a crystal tube was filled with xenon gas. Then, by means of a high explosive a strong shock wave was launched. The shock wave propagated along the tube. The radiation temperature observed was as high as 46,000C. Through the experiments, we could obtain various dependencies of physical quantities under different conditions. In the end, we could develop a laser system without using the crystal tube. The laser was further improved by changing the configuration of the surrounding explosives. As a result, laser energy of 100 kJ was successfully extracted at the quantum efficiency of about 100%.

We also studied other physical aspects of ion-laser such as beam divergence. We measured the level of optical inhomogenuity in active medium, and found it to be of order of 10^{-6}. So we could estimate the divergence of ion-laser using a plane resonator to be approximately 10^{-2} rad. Exactly, this number coincided with one obtained by experiments. Also, we invented another resonator, and could get beam divergence of up to 300 mrad. The results were obtained by using phase conjugation technique in stimulated Brillouin scattering (SBS) process, which was developed at Levedev institute in 1972. Beam divergence of 20 mrad was then achieved utilizing SBS with a kino-form optics. Thus, in the 1970s, we could understand clearly the behavior of ion-laser.

Early 1960s, nuclear physicists at Arzamas-16 noticed the possibility of compressing deuterium-tritium (D-T) mixture by irradiating laser on it. Academician Sakharov then reported in 1961 that D-T plasma can be compressed to a high density nearly with Fermi degeneracy by means of laser ablation. Of course, Basov and Krohkin also contributed to the study, and their work was published in 1964. In 1968, a few of our scientists theoretically considered indirectly driven targets for the first time. This type of target is now investigated world wide. In 1973, we began to develop ion-laser for nuclear fusion research. We have now two facilities, Iskra 4 and Iskra 5, which were put in operation in 1979 and 1989, respectively. The diameter of final amplifier of Iskra 5 is 70 cm. In the target chamber, the laser light can be focused up to the spot diameter of 10 mm.

Furthermore, I would like to talk shortly about one of the latest results. This shows calculated convergence ratios when neutrons are generated. It then turns out to be possible to attain volume compression by a factor of about 2,000. The difference in neutron yields between 1-D numerical calculations and experiments were approximately by a factor 2. By taking mixing of the D-T fuel and the glass as the pusher material into account, we could obtain even better description of the experiments. Currently, we are designing a new neodium glass laser facility with an output energy of about 500 kJ at first harmonics. Of course, it is necessary to construct more powerful laser system for ignition and high gain, but for various reasons, it is presently impossible. Now, we are specifying the detailed design such as laser energy, size of a beamlet, and total number of beams. At the moment we suppose the number of beams of 108-216. Meanwhile, we also try to design even larger facility with an energy of 10 MJ together with Livermore scientists. Now we think it possible to obtain 70 kJ from a single beam by optimizing the details. Then, energy output of 10 MJ may be achieved with 144 such beams.

Thus, at Arzamas-16, we have sufficient programs in laser-plasma physics for the next 10-15 years on Iskra 5 facility.

MICHAEL H. KEY
1997

THE EDWARD TELLER MEDAL LECTURE: HIGH INTENSITY LASERS AND THE ROAD TO IGNITION

M. H. Key

Lawrence Livermore National Laboratory
Livermore, California 94551

Abstract. There has been much progress in the development of high intensity lasers and in the science of laser driven inertially confined fusion such that ignition is now a near term prospect. This lecture reviews the field with particular emphasis on areas of my own involvement.

INTRODUCTION

The development of high intensity lasers and the evolution of ideas and experiments for laser driven inertially confined fusion are twin themes with which my scientific interests have been entwined for three decades. During that time the concept of ignition of inertially confined fusion with lasers has evolved from early suggestions that kilojoule scale of laser energy might be sufficient to a mature and accurate understanding that a megajoule laser is required for central spark ignition in indirectly driven implosions (1).

When I worked from 1966 at Queens University Belfast and from 1976 to 1996 at a central facility for academic users of high power lasers which was established in 1976 at the Rutherford Appleton Laboratory (RAL), the scale of laser energy compatible with the level of resources in those academic programs was limited and at RAL we could not go beyond the few kilojoule range. In the mid 1980s however new ideas in laser physics opened up the possibility of ultra-high intensity lasers of moderate energy which we exploited vigorously at RAL from 1989 onwards. Outstanding work by colleagues Chris Edwards and Colin Danson and Mick Shaw and Graham Hirst respectively contributed to two significant new laser facilities at RAL. The first was the adaptation of a beam line of the multi beam Nd glass laser system Vulcan to operate with chirped pulse amplification (CPA) which boosted the individual beam power from 0.5 to 40 TW (2). The second was development of very novel high intensity KrF CPA (3) and Raman (4) laser systems Sprite then Titania. With these lasers RAL scientists and facility user teams from UK universities and collaborators from overseas conducted some of the earliest experiments at more than 10^{19} Wcm^{-2}.

My most recent task since joining the Lawrence Livermore National Laboratory (LLNL) in October 1996 is to guide experiments designed to evaluate the new concept of fast ignition. I am fortunate in working at LLNL with Max Tabak and

CP406, *Laser Interaction and Related Plasma Phenomena*: 13th International Conference,
edited by G. H. Miley and E. M. Campbell

colleagues who originated the fast ignition scheme (5) and to have access to a petawatt CPA beam at the Nova laser developed by Mike Perry and colleagues (6), which is by an order of magnitude the most powerful laser beam available today.

FUSION PERSPECTIVE

After two decades of evolution, ignition of inertially confined fusion (ICF) has become the near term goal of projects in the USA and France which are based on building megajoule laser facilities. There is a high level of confidence in achieving ignition with indirect drive and research into ignition by direct drive has reached a mature level with the new megajoule lasers configured to include direct drive as an option. Gain up to ten fold in fusion energy relative to laser energy is expected, as illustrated in Fig. 1 showing the indirect drive point design for the US National Ignition Facility (NIF). The relative status of progress in magnetic fusion energy research (MFE) and ICF has been changed significantly by these developments as is now expected that ICF ignition will be obtained before a self sustaining burn in MFE.

A limitation in the ICF work is that the expected gain in indirectly driven targets is too low for realization of an energy source by inertial fusion IFE (1). For a laser driver with typically less than 10% efficiency, gain in excess of 100 is required. Moreover even with further optimization, the isobaric central spark ignition used in both indirect and directly driven implosions has a limiting gain (7) for a realistic 8% hydrodynamic efficiency and compression adiabat ratio of 2, which is less than 50 for circa 1 MJ driver energy as illustrated in Fig. 1. Laser driven IFE does not therefore appear feasible at all with the isobaric scheme. An ion beam driver of higher efficiency is at present the only IFE option which could utilize the lower gain of implosions with the isobaric central spark.

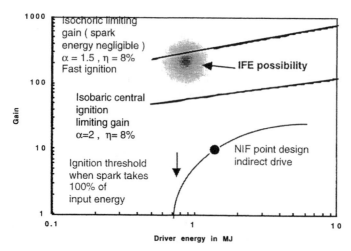

FIGURE 1.

With short pulse high intensity lasers came the new fast ignition scheme which is based irradiation in a single small focal spot of a pre-compressed target causing ignition under isochoric conditions for which the limiting gain exceeds 200 for 1 MJ driver energy. Although this new concept is relatively untested there is great interest in it as an alternative route to ignition because it opens up the possibility of IFE with lasers as a long term goal beyond the ignition facilities such as NIF.

SOME MILESTONES

The road to laser driven ICF ignition has been a long one on which some milestones can be identified. The first laser was demonstrated in 1960. The first laser initiated (but later understood to be non-thermonuclear) fusion reaction was reported in 1968 at the Lebedev Institute in the USSR. Much early research into the possibility of ICF was not in the public domain but in 1972 the principle of isobaric ignition in laser driven implosions was set our in a seminal publication by Nuckolls, Wood, Thiessen, and Zimmerman(8). Shortly after that in 1974 the KMS fusion Laboratory in the USA was the first to demonstrate thermonuclear fusion in a laser driven implosion (9). The use of indirect drive by thermal x-rays was pioneered behind a screen of classification at LLNL with early experiments from 1975 and continuous development up to the current NIF project. Declassification has recently opened up most of this area to wider involvement (1). Along the way a significant step was the discovery of laser plasma instabilities causing damaging preheating of the targets by energetic electrons leading to a vital change to the UV 3rd harmonic rather than the IR. fundamental frequency of the Nd glass laser for indirect drive . A major requirement in ICF is to achieve high density in order to reduce the critical mass and energy for ignition and in 1990 work at the Japanese Osaka University was the first to demonstrate a relevant density of 600g/cc in imploded polymer capsules using direct drive (10). Many other significant steps could also be identified from numerous laboratories world wide were this brief summary to be extended.

My personal milestones overlap in several ways with the road to laser driven ICF. My post graduate work at Imperial College in the early 1960's led me to build one of the first Q switched lasers and a theta pinch plasma in order to investigate the then new method of plasma diagnosis by Thomson scattering. As a post doctoral researcher in 1966 I began to study plasmas produced by Q switched laser radiation. Moving in 1966 as a newly appointed faculty member to Queens University Belfast I built up a Nd glass laser facility which became the most powerful in the UK universities and used it to study x-ray emission from laser produced plasmas, to develop an x-ray streak camera and related diagnostic capabilities and to conduct early research into x-ray lasers. In 1976 when it was agreed to establish at the Rutherford Laboratory new Central Facility for UK university work with high power lasers, I moved there to coordinate the scientific work of the new project and in 1983 I became Head of the Facility. In the next decade at RAL we further developed the Nd glass laser named Vulcan and a KrF laser facility which we called Sprite. I worked also in collaboration with university colleagues on problems motivated by interest in fusion, developing x-ray backlighting to study the dynamics of implosions (11) and hydrodynamic instabilities of laser accelerated targets (12). We also introduced new x-ray spectroscopy methods with streak time resolution (13) and space resolution for diagnostics such as preheat by hot electrons from K alpha fluorescence (14) and measurements of mass ablation rate (15). X-ray laser research became a substantial topic a highlight being demonstration of saturated x-ray laser operation for the first time (16).

A significant reorientation of the RAL work began in 1989 with our decisions to emphasize high intensity physics by adapting Vulcan for CPA operation at up to 40 TW. 0.8 ps power and building a larger scale KrF laser Titania for both CPA and Raman laser operation. The KrF work is of a pioneering nature and the Raman system is particularly novel and unique. It offers near diffraction limited beams of up to 100 J energy and down to 20 ps pulse length. In parallel with these developments I was invited to set up a university research group at Oxford University and with colleagues and students from Oxford studied topics in high intensity physics including optical field ionized plasmas (17), XUV high harmonics (18) and the application of x-ray lasers to measure laser imprinted hydrodynamic perturbations of importance in direct drive ICF (19).

RECENT HIGH INTENSITY SCIENCE AT RAL

The capability for experiments at up to 10^{19} Wcm^{-2} at RAL has been important for the wider programs of facility users and much new science has been accomplished through and annual rate of up to 1000 high intensity shots over the last three years. Examples include measurements of relativistic self focusing (20) and high energy ion emission (21) and accelerated electrons (22).

Members of my Oxford University research group also contributed as users of the RAL lasers as the following examples illustrate . XUV high harmonic generation has interesting overlaps with x-ray laser research and offers some new possibilities for generating ultra-short pulses of high brightness. Two aspects here were of particular interest to us in connection with the Sprite/Titania laser facilities,which generate UV light unlike most of the lasers used for research into harmonic generation. The first was the possibility that harmonic conversion from ions rather than atoms would lead to shorter wavelength harmonics and the second was the anticipation from earlier work at wavelengths 1.05 and 0.53 micron, that a UV laser would give a higher conversion efficiency to harmonics of a given wavelength. In the event both these possibilities were successfully demonstrated, the first using the Sprite CPA laser leading to the shortest wavelength XUV harmonic with the 6.72 nm 37th harmonic of 248.6 nm from He$^+$ ions (18) and the second using the CPA beam of Titania with production of more than 10^{-6} conversion into the 7th harmonic (23).

A most recent development was the pushing forward of the short wavelength limit at which saturated x-ray laser action has been observed using six beams of Vulcan operating with 70 ps pulse duration to irradiate a double slab target in two in line 100 micron wide line foci. The target material was Sm and the laser action on the J=0-1 line of the Nickel-like ionization stage was at 7.3 nm with a power of 6 MW (24) illustrated in Fig. 2 .

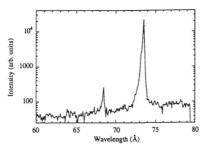

FIGURE 2. The spectrum of the Sm X-ray laser.

Continuing the theme of harmonics and short wavelengths the CPA beam of Vulcan was shown to produce an extended series of both odd and even harmonics when solid slab targets were irradiated at 10^{19} Wcm^{-2} (25). The mechanism is the anharmonic motion of free electrons driven in and out across the critical density surface by the axial component of the electric field at the turning point of obliquely incident p polarized light. The novel feature of the results was that much shorter wavelength than in other experiments was observed down to the 75th harmonic of 1.05 micron and this was directly related to the high intensity. In an interesting follow up experiment the Doppler shift of the 4th harmonic was studied and it was shown that the extreme light pressure caused the critical density surface to recede at up to 1.5% of the velocity of light (26) as shown in Fig. 3, a process of interest for fast ignition which is discussed later, since it causes hole boring and steepening of the density profile.

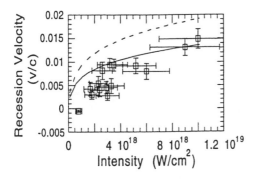

FIGURE 3. Recession velocity inferred from Doppler shifted 4th harmonic emission.

7

DIRECTLY DRIVEN ICF

The majority of investment of research effort in ICF has gone into indirect drive but direct drive has remained of significant interest both because it is conceptually somewhat simpler and because its critical issues are different from indirect drive so that it provides a fall back option should unforeseen difficulties arise in indirect drive.

Perhaps the most critical issue in direct drive is the seeding of hydrodynamic instability in the implosion by the speckle pattern of the laser light driving the target. This imprinting process adds significantly to the initial perturbations of the spherical surface residual from its fabrication. The amplitudes of the perturbations is typically on the scale of a few tens of nanometers but the growth factors for Rayleigh Taylor instability are larger in direct than in indirect drive and the amplitude and frequency spectrum of the perturbations is a critical issue.

A natural extension of my interest in radiography and development of x-ray lasers was to apply the x-ray laser as a radiographic source to measure these small perturbations (19) as part of a broader effort to evaluate the direct drive fusion option which I embarked upon during a sabbatical year spent at LLNL in 1994/95.

The XUV laser offers high brightness and short pulse and together with normal incidence XUV optics, high spatial resolution in radiography. The high brightness enables measurements through targets with very large attenuation giving significant changes in transmission for small fractional changes in thickness.

This work began with colleagues at LLNL using the Nova laser (27) and was continued at the Vulcan facility which is better adapted for the purpose. Figure 4 illustrates 19.6 nm Ge laser radiographs of laser driven thin Al foils for three kinds of optical smoothing used in direct drive, the static speckle pattern of a random phase plate RPP, the one dimensionally fluctuating speckle of smoothing by spectral dispersion SSD and two dimensionally fluctuating speckle of induced spatial incoherence ISI (28). High quality data are obtained showing the thickness

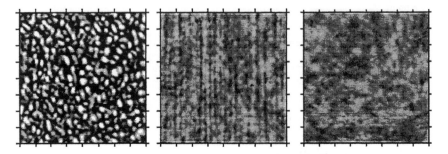

FIGURE 4. Radiographs showing 200 micron regions of 2 micron thick Al foil driven at $5 \ 10^{12}$ Wcm^{-2} recorded by XUV laser radiography for RPP. SSD and ISI smoothing respectively from left to right.

perturbations of the targets for amplitudes as low as 30 nm. In related work a single optical mode (produced as a sinusoidal fringe pattern using a double slit aperture in the laser beam before the focusing lens) has been used to study a physically simpler system to allow more precise comparison with theory.

FAST IGNITION

At LLNL there is a commitment to evaluate fast ignition both for its long term potential in IFE and for its nearer term interest as a possible adaptation of the NIF facility. The 1 kJ,1 PW beam line at Nova was built primarily for this task.

The relevant parameter range for the physics of fast ignition can be assessed from the fairly well established requirements of the ignition spark. Isochoric ignition needs some what higher temperature (kT =15 keV) and higher density radius product (ρr = 0.6 g/cc) than the 0.3 g/cm^2, 5 keV of isobaric ignition, because of the dissipation of energy in blast wave expansion of the spark. If 200 g/cc is assumed for the compressed target the required radius is 30 micron and the thermal energy is 35 kJ. If laser energy were converted to electrons and these were absorbed in the spark volume with a net 40% efficiency the laser energy would be 90 kJ. Well collimated electrons and efficient absorption are required to achieve 40% net coupling. There will inevitably be some separation between the dense core and the laser absorption region which is estimated as of order one focal spot diameter and there must be some related increase in area from the laser focal spot to the heated spark which for plausibly well collimated electrons, is at least a factor of two. The pulse duration must be less than the inertial confinement time which is 30 ps in this example and if we therefore specify a 10 ps pulse, the necessary laser intensity is 6 10^{20} Wcm^{-2}.

The mean energy E of the electrons from PIC modeling is similar to the ponderomotive potential (5) which scales as $(I\lambda^2)^{1/2}$ in the relativistic limit. For the example considered assuming 1 micron wavelength, E would be 6 MeV. This illustrates a problem in that we require the range of the electrons to match the ρr of the ignition spark. An energy of 1 MeV is indicated unless collective behavior shortens the range. A solution here may be to use a shorter laser wavelength and in this example the required wavelength would be less than 0.2 micron.

A further interesting and often quoted possibility in fast ignition is to reduce the ignition energy by scaling up the density ρ. There is a ρ^{-2} scaling of energy and a ρ^{-1} scaling of power with pulse length also scaling as ρ^{-1}. The apparent benefit is obvious but the required laser intensity scales as ρ and exacerbates the problem of the electron energy.

A short wavelength is attractive also through reduction of the length of plasma with density above a few percent of critical because difficulties of channeling through the plasma increase with the length involved.

Experiments using the petawatt laser and detailed modeling are being carried out at

LLNL to study and evaluate fast ignition. This work at a wavelength of 1.05 micron, is emphasizing the heating effect of the fast electrons and scaling behaviors. Results are preliminary and not yet published but early indications are encouraging and near keV temperature at near solid density has been inferred from x-ray spectroscopy and DD fusion yield from buried layers in plane targets.

It is also apparent that while fast ignition requires short pulses it does not require ultra-short pulses and CPA lasers are not then so advantageous. The KrF Raman laser can generate a near diffraction limited beam at 268 nm in pulses as short as 30 ps and its intensity limit in the beam is about 30 GW/cm^2 or 0.9 J cm^{-2}. A CPA laser with metallic gratings is limited to 0.2 J/cm^2 and with dielectric gratings and pulses up to 30 ps this figure may be increased to about 1 Jcm^{-2}. The two technologies therefore appear to have similar capabilities of energy delivery per area of beam.

It is much too early to draw any conclusions on the viability of fast ignition and on the required wavelength but it would be an interesting closure of loops were I to find as a result of my move to LLNL that fast ignition needs the KrF laser technology which we pioneered at RAL!

ACKNOWLEDGMENTS

I am grateful to many colleagues at RAL, Oxford University, other UK universities, LLNL and in other laboratories outside the UK for their contributions to the research and development activities reviewed here. Experimental science is in essence team work and I have been fortunate in the quality of the team members with whom I have worked.

Work performed under the auspices of the U.S. Department of Energy by the Lawrence Livermore National Laboratory under Contract No. W-7405-ENG-48.

REFERENCES

1. Lindl, J. D., "The evolution towards indirect drive and two decades of progress towards ICF ignition and burn," Laser Interaction and Related Plasma Phenomena, AIP Conf. Proc. 318 , p. 635, (1993).
2. Danson, C. N., Barzanti, L., Chang, Z., Damerell, A., Edwards, C. B., Hancock, S., Hutchinson, M. H. R., Key, M. H., Luan, S., Mahadeo, R., Mercer, I. P., Norreys, P., Pepler, D. A., Rodkiss, D. A., Ross, I. N., Smith, M. A., Smith, R. A., Taday, P. F., Toner, W. T., Wigmore, K., Winstone, T. B., Wyatt, R. W., "High contrast multi-terawatt pulse generation using chirped pulse amplification on the VULCAN laser facility," Opt. Comm., **103**, 392, (1993).
3. Ross, I. N., Damerell, A. R., Dival, E. J., Evans, J., Hirst, G. J., Hooker, C. J., Key, M. H., Lister, J. M. D., Osvay, K., M. J. Shaw, "A 1 TW KrF laser using chirped pulse amplification," Opt. Comm., **109**, 288 (1994).

4. Ross, I. N., Shaw, M. J., Hooker, C. J., Key, M. H., Harvey, E. C., Lister, J. M. D., Andrews, J. E., Hirst, G. J., Rogers, P. A., "A high performance excimer pumped Raman laser," Opt. Comm., **78**, 262, (1990).

5. Tabak, M., Hammer, J., Glinsky, M. E., Kruer, W. L., Wilks, S. C., Woodworth, J., Campbell, E. M., Perry, M. D., "Ignition and high gain with ultra-powerful lasers," Phys. Plasmas **1**, 1626 (1994).

6. Perry, M. D., and Morou, G., "Terawatt to petawatt subpiocosecond lasers," Science **264**, 917, (1994).

7. Atzeni, S., "Thermonuclear burn performance of volume ignited and centrally ignited bare deuterium tritium microspheres," Jpn J. Appl. Phys. **34**, 1980 (1995).

8. Nuckolls, J., Wood, L., Thiessen, A., Zimmerman, G., "Laser compression of matter to superhigh densities: thermonuclear (CTR) applications," Nature **239**, 139 (1972).

9. Campbell, P. M., Charatis, G. G., Montry, R., "Laser driven compression of glass microspheres Phys. Rev. Lett. **34**, 74, (1975).

10. Azechi, H., Jitsuno, T., Kanabe, T., Katayama, M., Mima, K., Miyanaga, M., Nakai, M., Nakai, S., Nakaishi, H., Nakatsuka, M., Nishiguchi, A., Norreys, P. A., Setsuhara, Y., Takagi, M., Yamanaka, M., Yamanaka, C., "High-density compression experiments at ILE, Osaka," Lasers and Particle Beams **9**, 193, (1991).

11. Key, M. H., Lewis, C. L. S., Lunney, J. G., Moore, A., Hall, T. A., Evans, R. G., "Pulsed x-ray shadowgraphy of dense cool laser imploded plasma," Phys. Rev. Letts. **41**, 1467 (1978).

12. Rumsby, P. T., Key, M. H., Hooker, C. J., Cole, A. J., Kilkenny, J. D., Evans, R. G., "Measurement of Rayleigh-Taylor instability in a laser accelerated target," Nature **299**, 329 (1982).

13. Key, M. H., Lewis, C. L. S., Lunney, J. G., Moore, A., Ward, J. M., Thareja, R.K., "Time resolved x-ray spectroscopy of laser produced plasmas," Phys. Rev. Lett. **44**, 1669 (1980).

14. Hares, J. D., Kilkenny, J. D., Key, M. H., Lunney, J. G.,"Measurement of fast electron energy spectra and preheating in laser irradiated targets," Phys. Rev. Letts. **42**, 1216 (1979).

15. Key, M. H., Toner, W. T., Goldsack,, T. J., Veats, S. A., Kilkenny, J. D., Cunningham, B. J., Lewis, C. L. S., "A study of laser ablation by laser irradiation of plane targets at wavelengths 1.06, 0.53 and 0.35 micron," Phys. Fluids **26**, 7, 2011 (1983).

16. Carillon, A., Chen, H. Z., Dhez, P., Dwivedi, L., Jaegle, P., Jamelot, G., Zhang, J., Key, M. H., Kidd, A., Klisnick, A., Kodama, R., Krishnan, J., Lewis, C. L. S., Neely, D., Norreys, P., Oneill, D. M., Pert, G. J., Ramsden, S. A., Raucourt, J. P., Tallents, G. J., Uhomoibhi, J., "Saturated and near diffraction limited operation of an XUV laser at 23.6 nm," Phys. Rev. Lett., **68**, 2917, (1992).

17. Blyth, W. J., Preston, S. G., Offenberger, A. A., Key, M. H., Wark, J. S., Najmudin, Z., Modena, A., Djaoui, A., Dangor, A. E., "Plasma temperature in optical field ionisation of gases by intense ultra-short pulses of ultraviolet radiation," Phys. Rev. Lett. **74**, 554, (1995).

18. Preston, S. G., Sanpera, A., Zepf, V., Blyth, W. J., Smith, C. G., Burnett, K., M. H., Key, Wark, J. S., Neely, D., Offenberger, A. A., "High order harmonics of 248.6 nm KrF laser from helium and neon ions," Phys. Rev. A, **53**, R31, (1996).

19. Key, M. H., Barbee Jr., T. W., DaSilva, L. B., Glendinning, S. G., Kalantar, D. H., Rose, S. J., Weber, S. V., "New plasma diagnostic possibilities from radiography with XUV lasers," J Quant . Spectr. and Radiative Transfer, **54**, 221, (1995).

20. Borghesi, M., MacKinnon, A. J., Baringer, L., Gaillard, R., Gizzi, L. A., Meyer, C., Willi, O., Pukhov, A., Meyer ter Vehn, J., "Relativistic channelling of a picosecond laser pulse in a near critical preformed plasma," Phys. Rev. Letts. **78**, 879 (1997).

21. Fews, A. P., Norreys, P. A., Beg, F. N., Bell, A. R., Dangor, A. E., Danson, C. N., Lee, P., Rose, S. J., "Plasma ion emission from high intensity picosecond laser pulse interactioins with solid targets," Phys. Rev. Letts. **73**, 1801 (1994).

22. Modena, A., Najmudin, Z., Dangor, A. E., Clayton, C. E., Marsch, K. A., Joshi, C., "Electron acceleration from the breaking of relativistic plasma waves," Nature **377**, 606 (1995).

23. Preston, S. G., Chambers, D. M., Marjoribanks, R. S., Norreys, P. A., Neely, Zepf, M., Zhang, J., Key, M. H., Wark, J. S., "A krypton fluoride laser source of bright, extreme-ultraviolet harmonic radiation," Phys. Rev. A (submitted).

24. Zhang, J., McPhee, A. G., Lin, J., Wolfrum, E., Smith, R., Danson, C., Key, M. H., Lewis, C. L .S., Neely, D., Nilsen, J., Pert, G. J., Tallents, G. J., Wark, J. S., "A saturated x-ray laser beam at 7 nm," Science **276**, 1097 (1997).

25. Norreys, P. A., Zepf, M., Moustaizis, S., Fews, A. P., Zhang, J., Lee, P., Bakarezos, M., Danson, C. N., Dyson, A., Gibbon, P., Loukakos, P., Neely, D., Walsh, F. N., Wark, J. S., Dangor, A. E., "Efficient extreme UV harmonics generated from picosecond laser pulse interactionswith solid targets," Phys. Rev. Lett. **76**, 1832 (1996).

26. Zepf, M., Castro-Colin, M., Chambers, D., Dangor, A. E., Danson, C. N., Dyson, A., Fews, A. P., Gibbon, P., Key, M. H., Lee, P., Moustaizis, S., Neely, D., Norreys, P. A., Preston, S. G., Wark, J. S., Zhang, J., "Measurement of the hole boring velocity from Doppler shifted harmonic emmission from solid targets," Phys. Plasmas, **3**, 3242, (1997).

27. Kalantar, D. H., Key, M. H., DaSilva, L. B., Glendinning, S. G., Knauer, J. P., Remington, B. A., Weber, F., Weber, S. V., "Measurements of 0.35 micron laser imprint in a driven Si foil by XUV laser radiography," Phys. Rev. Letts. **76**, 3574, (1996).

28. Kalantar, D. H., Da Silva, L. B., Demir, A., Glendinning, S. G., Key, M. H., Kim, N, S., Knauer, J. P., Lewis, C. L .S., Lin, J, Neely, D., MacPhee, A., Remington, B. A., Smith, R., Tallents, G. J., Wark, J. S., Warwick, J., F. Weber, Weber, S. V., Wolfrum, E., Zhang, J., "XUV probing of laser imprint in a thin foil using an x-ray laser backlighter," Rev .Sci. Instr. **68**, 802 (1997).

JÜRGEN MEYER-TER-VEHN
1997

ICF Related Research at MPQ

J. Meyer-ter-Vehn

Max-Planck-Institut für Quantenoptik, D-85748 Garching, Germany

Abstract. A personal account is given on research at the Max-Planck-Institut für Quantenotik (MPQ) related to inertial confinement fusion (ICF).

THE YEARS 1979 - 1983.

In accepting the Edward Teller Award, I would like to share the prize with the whole laser-plasma group at MPQ. Without the stimulation of this group, led by Siegbert Witkowski, my personal theoretical contributions would have been impossible. In particular, the experiments initiated by Richard Sigel have been a continuous challenge for me. When I joint the group in 1979, it was in connection with heavy ion beams and their potential as ICF driver. **Rudolf** Bock from GSI Darmstadt had been the first in Germany to realize that there is emerging a major new option for research at a heavy ion lab like GSI, and he was starting an exploratory programme. This was in the wake of the Nature article by Nuckolls et al. in 1972 [1], which had set time zero for open ICF research.

With no experience in this field, I bought the book of Zeldovich and Raizer [2] and found very basic and fascinating physics I had never considered before. The papers by Ray Kidder [3] were revelations of how to understand the essentials in analytic terms. My first paper on ICF [4], which then became quite popular and probably earned me the present award, contained just a slight modification of Kidder's work. Assuming fuel pressure to be uniform at stagnation (rather than density), I could reproduce the Livermore (simulation-based) gain predictions within a simple model quite accurately, as seen in Fig. 1. Beside the physical insight, it provides important scaling relations for general studies. We used them to define the working point in the HIBALL study, a design of a heavy ion driven ICF reactor, which was published in 1983 [5].

CP406, *Laser Interaction and Related Plasma Phenomena*: 13th International Conference,
edited by G. H. Miley and E. M. Campbell

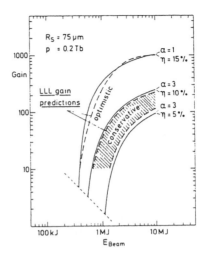

FIGURE 1: Isobaric gain model fitted to Livermore gain curves; parameters: p fuel pressure at ignition, R_S radius of ignition spark, $\alpha = p/p_{deg}$ fuel isentrope, η fraction of beam energy in igniting fuel.

Spherical implosion for achieving 1500 times fuel compression at pressures of 200 Gbar is the central problem for ICF gas dynamics. In this respect, the paper of Gottfried Guderley on converging shocks, published 1942 in wartime Germany, became of particular importance for me [6]. From Stephen Coggeshall I learned that it is Lie group symmetry which is at the basis of these amazing similarity solutions [7]. After years of search, I found that Guderley was alive, living in Dayton/Ohio. In a letter he told me that the motivation behind his 1942 paper was pure science, not related to weapons.

INDIRECT DRIVE

Laser plasma experiments at MPQ had been plagued by non-uniform energy deposition on targets for a long time, when Sigel started to explore indirect drive in 1983. The first step was to shine laser light into a gold cavity to produce Planckian radiation. A key question was what temperature could be reached, and it was Ricardo Pakula, one of Sigel's students, who first came out with an answer in terms of a similarity solution, describing the X-ray diffusion into the cavity wall [8]. It provided hohlraum scaling relations that describe experiments quite well.

Pushing for higher temperatures, Sigel initiated joint experiments with the Osaka group using their GEKKO XII laser. Milestones of this work were the experimental demonstration of radiative heat waves propagating through gold walls with sharp fronts and applications of hohlraum X-rays to generate very uniform shock waves [9]. This research line culminated in the hands of Thorsten Loewer, Sigel's former student, who developed the Labyrinth hohlraum for single beam heating with the MPQ ASTERIX laser [10]. With 500 J input pulses, he could demonstrate 140 eV temperatures and uniform shock up to 20 Mbar *free of preheat* [11].

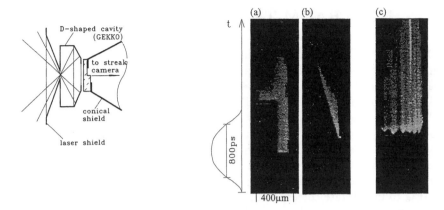

FIGURE 2: (lhs) D-shaped hohlraum used in ILE shock experiments [11], (rhs) streak records of shock signals from (a) stepped gold sample (notice uniform fronts!), (b) gold wedge (notice constant shock velocity!), (c) flat sample directly driven by laser, for comparison.

Cross-sections for soft X-ray absorption in dense plasma govern the radiation transport, and Klaus Eidmann at MPQ was among those pioneering transmission spectroscopy to measure opacities. Precise data were obtained for elements from Be to Au at densities of 0.01 -0.10 g/cm^3 and temperatures of about 20 - 30 eV [12]. With no high-Z opacities available in the literature, we also started our own opacity calculations, and results are contained in the final report of the Opacity Workshop Work-OpIII, held at MPQ in 1994 [13].

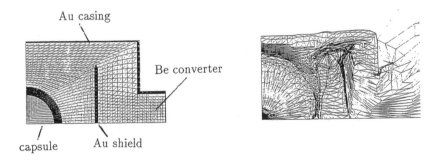

FIGURE 3: MULTI-2D simulation of a hohlraum target for heavy ion fusion; initial configuration (lhs), state at end of implosion (rhs).

HOHLRAUM TARGET SIMULATIONS

Obviously, there was a need for numerical simulations of radiation hydrody-namics, and it was Rafael Ramis from UPM Madrid who developed the corresponding MULTI code [14]. He visited MPQ regularly since 1986. MULTI-1D, a 1D-hydro code for multigroup radiation transport , has become an everyday tool for analyzing experiments at MPQ. The 2D version MULTI-2D, though still under development, is already used for hohlraum target design (see Fig. 3 and ref. [15]). Regarding heavy ion fusion, it had become evident after the HIBALL study that indirect drive is needed to solve the symmetry problem. Together with Masakatsu Murakami from ILE Osaka, we started in 1989 at MPQ to investigate corresponding hohlraum configurations by MULTI-1D simulations and checking symmetry by means of 2D-viewfactor treatment [16]. Meanwhile a number of researchers including Stefano Atzeni, Micheal Basko, Javier Honrubia, Joachim Maruhn, Julio Ramirez, Rafael Ramis, and some students and post-docs have joint this effort and have come out with significantly refined studies of different designs [17].

FAST IGNITION

A major restructuring of the MPQ laser-plasma group occured in 1993 with the retirement of S. Witkowski and the decision to close down the ASTERIX laser. Klaus Witte took over responsibility for the new CPA laser facility ATLAS at MPQ. At present, the goal is to increase the 1TW/150fs pulses available now to 10-100TW with intensities up to 10^{20}W/cm^2. Within the ICF keep-in-touch activity funded by EURATOM, the ICF related work at MPQ will now focus on physics of fast ignition of fusion targets.

On the theory side, this opens up the new field of relativistic laser-plasma physics with a large number of new and fascinating options. At intensities beyond 10^{18}W/cm^2, electrons are accelerated to MeV energies and mainly in forward direction. This implies radically different interaction dynamics and requires kinetic simulation. With the development of a three-dimensional particle-in-cell (PIC) code solving full Maxwell's equations and relativistic particle dynamics, MPQ has taken a leading step in this direction. The code has been developed at MPQ by Alexander Pukhov, a guest from MIFT Moscow. Figure 5 shows a 3D-PIC simulation of a 3×10^{19}W/cm^2 pulse propagating in near-critical plasma. It demonstrates formation of a light channel $1 - 2\mu$m wide. A relativistic electron beam concomitant with the light pinches in its self-generated magnetic field of about 100 Mega-Gauss and guides the light through relativistic effects [18]. Relativistic self-focussing has been demonstrated on ATLAS [19]. Laser hole boring into overdense plasma relevant for fast ignition is presently under investigation.

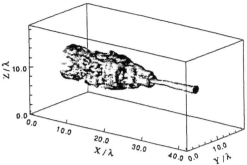 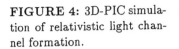

FIGURE 4: 3D-PIC simulation of relativistic light channel formation.

REFERENCES

1. Nuckolls, J., et al., Nature **239**, 139 (1972).
2. Zeldovich, Ya.B., and Raizer, Yu.P., *Physics of Shock Waves and High Temperature Hydrodynamic Phenomena*, New York: Academic Press, 1967.
3. Kidder, R.E., *Nuclear Fusion* **16**, 405 (1976).
4. Meyer-ter-Vehn, J., *Nucl. Fusion* **22**, 561 (1982).
5. Böhne, D., et al., *Nucl.Engineering and Design* **73**, 195 (1982).
6. Guderley, G., *Luftfahrtforschung* **19**, 302 (1942).
7. Coggeshall, S.V., and Meyer-ter-Vehn, J., *J. Math. Phys.* **33**, 3585 (1992).
8. Pakula, R., and Sigel, R., *Phys. Fluids* **28**, 232 (1985), and **29**, 1340 (1986).
9. Sigel, R., in Laser-Plasma Interactions, ed. M.B. Hooper, SUSSP Publ., Edinburgh, vol. 4 (1989), pp. 53-87, and vol.5 (1995), pp. 79-104.
10. Baumhacker, H., Brederlow, G., Fill, E., Volk, R., Witkowski, S., Witte, K., *Appl. Physics* **B61**, 325 (1995).
11. Löwer, Th., Nishimura, H., et al., *Phys. Rev. Lett.* **72**, 3186 (1994), and Löwer, Th., Basko, M., (1997) to be published.
12. Winhart, G., Eidmann, K., Iglesias, C.A., Bar-Shalom, A., Phys.Rev. E53, R1332 (1996).
13. Rickert, A., Eidmann, K., Meyer-ter-Vehn, J., *Final Report of Third Int. Opacity Code & Code Comparison Study*, Max-Planck-Institut für Quantenoptik, Garching, Report MPQ204 (Aug.1995).
14. Ramis, R., et al., *Comp. Phys. Comm.* **49**, 475 (1988) and Report MPQ174 (Aug.1992).
15. Meyer-ter-Vehn, J., Ramirez, J., Ramis, R., *Fus. Engin. Design* **32-33**, 585 (1996).
16. Murakami, M., and Meyer-ter-Vehn, J., *Nucl. Fusion* **31**, 1315 (1991).
17. Atzeni, S., et al., *Proc. 16th IAEA Fusion Energy Conf.*, Montreal, Canada, Oct. 7-11, 1996.
18. Pukhov, A., Meyer-ter-Vehn, J., *Phys. Rev. Lett.* **76**, 3975 (1996).
19. Fedosejevs, R., Wang, X.F., Tsakiris, G., submitted to *Phys.Rev.E* (May, 1997).

GUILLERMO VELARDE
1997

GV826-NEbis

THE LONG WAY TOWARDS INERTIAL FUSION ENERGY

Guillermo Velarde

In 1955 the first Geneva Conference was held in which two important events took place. Firstly, the announcement by President Eisenhower of the Program Atoms for Peace declassifying the information concerning nuclear fission reactors. Secondly, it was forecast that due to the research made on stellerators and magnetic mirrors, the first demo fusion facility would be in operation within ten years. This forecasting, as all of us know today, was a mistake. Forty years afterwards, we can say that probably the first Demo Reactor will be operative in some years more and I sincerely hope that it will be based on the inertial fusion concept.

Around 1950, Edward Teller established the bases of the Inertial Confinement Fusion at larger scale and smaller density that the proper ones corresponding to the Inertial Fusion Energy. Ten years afterwards, Academician Nicolai Basov and John Nuckolls suggested the use of laser for the confinement and ignition of micropellets of deuterium and tritium.

As many brilliant theories, the original idea was apparently simple. When a laser illuminates uniformly the surface of a deuterium-tritium micropellet, ablation of its surface generates a pressure wave that compresses and heats the micropellet up to the fusion temperature. From the beginning, it was observed that fluid instabilities caused by illumination or target non-uniformities, could result in the mixing between the pusher and the fuel which prevented the compression of the micropellet.

Several systems to uniformize the laser illumination were developed, among them the indirect-driven targets, the so-called holhraum, based in some ways on the H bombs. Therefore the nuclear countries classified from the very beginning the research on ICF, mainly about holhraum. Although necessary at first, this classification produced two negative effects. The first one was the prohibition for the American scientists to publish sensitive theoretical and experimental results, whereas scientists from Italy, Germany, Spain and specially Japan could publish their work without any restriction. Sometimes the works published by these scientists were already done by their American colleagues. This fact produced nuisances that affected negatively to the quiet environment and collaboration which are so important in any process of scientific research.

The other negative effect, still worst, was that many politicians and scientists in several countries thought that the research on inertial fusion was of military interest for the development of H bombs and that Inertial Fusion did not have any civilian application. This negative effect was specially bad in the European Union, in which pacifist groups have a strong influence in some countries of the Union. These countries have opposed from then that the European Union could have its own independent program on IFE.

In this confusing and complicated environment, in June 1988 the 19th ECLIM was held in Madrid. Relevant scientists and directors of different laboratories devoted to the ICF participated in this Conference. Professors Chiyoe Yamanaka, Erik Storm, Vladislav Rozanov, Heinrich Hora and myself took advantage of this opportunity to decide two important agreements. On the one hand to write the Madrid Manifesto which was signed by over 130 scientists. In this Manifesto it was said:

Recent research results in Inertial Confinement Fusion (ICF) have put to rest fundamental questions about the basic feasibility of achieving high gain ICF, and make it clear that there should be an aggressive program to design, build and operate ICF facilities to demonstrate high gain fusion in the laboratory... The time has arrived to begin to seriously seek a new age in the development of ICF. The laudable goal of the international ICF community is to use the fullest possible collaboration among nations in order to provide the technological benefits from fusion that will serve all humanity.

In the spirit of the 19th European Conference on Laser Interaction with Matter (ECLIM) held in Madrid, we urge the international community to take action now.

The second agreement taken during the 19th ECLIM was to create the **International Society for Inertial Fusion Energy** whose objective was to promote international collaboration and to propose to the UNESCO and to the IAEA the publication of introductory books on this subject, as well as to have international meetings where non-specialist scientists in inertial fusion could learn that IFE is a viable way to get an abundant and environmentally sound source of energy. This Society integrated initially over 150 scientists with a Board made by myself as President, and Academician Nicolai Basov, Prof. Chiyoé Yamanaka and Prof. Heinrich Hora as Vicepresidents. In 1990 Academician Robert Dautry and in 1991 Dr. John Nuckolls joined the Society as Vicepresidents. For not interfering in declassification programs, some colleagues suggested us to keep the Society in stand by for some years.

During the 19th ECLIM Conference, the directors of some European laboratories working in inertial fusion such as Caruso, Fabre, Key, Witkowsky and myself had meetings to study the viability of an Eurolaser Facility. We decided to submit the results of this study under the consideration of our respective governments. However, and due to politic opportunism, this great idea did not come out, although this was the starting point for international cooperation in Europe in the field.

Following the ideas of the International Society for Inertial Fusion Energy, the first step we identify was the training of the students in our Universities through special programs and doctorate studies in Inertial Fusion. To this respect, in 1989 Professors Martínez-Val, Ronen and myself began the preparation of a text book titled **Nuclear Fusion by Inertial Confinement. A Comprehensive Treatise**, with collaboration of several outstanding scientists in this field. The US DOE finally allowed the American scientists working in national laboratories to collaborate, after reaching an agreement for not to include anything related to indirect targets, which was in fact one third of the book. The book was published by CRC Publishing of Florida.

In the preface of the book we said: *This book is restricted to direct-drive because indirect drive involves several sensitive topics which are classified in some countries and are not frequently treated in the open bibliography. Moreover, not all the authors involved in the book could accept the inclusion of this subject in our pages, and we did not wish to induce a schism in our scientific community.*

The second step to be taken was to gradually convince other scientists whose work was not related to Inertial Fusion Energy, as well as to the politicians responsible for scientific development in the respective countries, that Inertial Fusion Energy was not only an alternative to MFE but probably the best way to follow in the energy of the future.

To begin with this step, in 1990 Academician Basov and I considered that we should have the support of an international organization such as the UNESCO or the IAEA. By that time, the General Director of UNESCO was, and still is, Professor Mayor Zaragoza, an outstanding Spanish biologist and former Minister of Education and Scientific Research in Spain. Professor Mayor Zaragoza was worried because of the energy problem and immediately understood and agreed with our proposal, suggesting that the UNESCO would sponsor with US$100,000 an international meeting with the representatives of scientific policy in several countries so as to make a feasibility study on IFE which could lead in a next future to the IFER, the Inertial Fusion Energy Reactor, in a similar way to ITER. Again, some of these nuclear countries considered that this study could interfere in the declassification process of IFE, and for this reason I decided to postpone everything until the declassification should be finished.

In 1992 the Herald Tribune and the New York Times made public that *The Federal Government... is beginning to declassify some of the most sensitive aspects of its design and to let American scientists publish them in scientific literature. The reason for this reversal is not internal policy considerations, the end of the cold war or the collapse of the Soviet Union as a military threat. Rather it is foreign competition. Scientists in Japan, Germany, Spain and Italy... have openly published the "secrets" for years. Continued secrecy for similar research in the United States was seen as stifling the exchange of ideas, inhibiting progress and limiting international cooperation. At times American scientists have been ordered not to attend meetings with foreign scientists, because they would have run the risk of discussing classified information. As a result, the Department of Energy, the keeper of the secrets, carried out one round of declassification in 1990, and says it is readying another.* After this event, the US DOE decide to declassify almost all IFE research.

As a result of this new policy, the US DOE allowed the American scientists to collaborate in the book **Energy from Inertial Fusion** published in 1995 by the IAEA in Vienna. In this book indirect-driven targets were included for the first time.

Inertial Fusion Energy in the European Union.

From 1993, access to large European laser facilities is possible via the Commission´s Training and Mobility of Researchers Programme, formerly known as Human Capital and Mobility Programme. This programme is totally independent of the Community Fusion Programme.

Access to defense oriented facilities as CEA-PHEBUS is autorised only if the proposed experiments are accepted by the CEA authorities. Currently, about 15% of the PHEBUS operation time is devoted to the experiments proposed by laboratories of the European Union. The CEA intends to offer the same kind of access to its future Large Mega Joule facility.

In 1995 the European Science and Technology Assembly, one official advisory body to the European Union, held a Working Party on Inertial Confinement Options to Controlled Nuclear Fusion to evaluate the current situation of scientific and technical development of IFE and, consequently, to propose the creation of an independent programme for inertial fusion. The Working Party was chaired by Juan Rojo, former Deputy Minister of Scientific Research of Spain, with the collaboration of Carlo Rubbia.

To carry out this study, the Working Party held a Topical Workshop in December 1995 in Abingdon, UK, with the participation of several scientists from Europe and the US. In the Recommendations and Report of the ESTA Working Party, it was requested from the European Union to establish immediately a modest programme on inertial fusion research, with an initial financial support running at around 10 percent of the total fusion budget, and to have as well a power producing inertial confinement reactor as its ultimate goal.

The Working Party is of the opinion that controlled nuclear fusion is in the research phase and recommends that an ICF component should be included in the EU programme. Even more, it recognized that Ignition can be attained by Inertial Confinement before than by Magnetic Confinement.

This ICF programme should be of moderate size and should be established by the Commission in consultation with the scientists involved. There should be an initial programme phase of 5 years duration.

The journal Nature of April 25, 1996 published the following: The main thrust of the report produced by the ESTA working party on Inertial Confinement Options to Controlled Nuclear Fusion is that the EU is making a mistake of putting all its eggs in one basket by spending almost all its annual funding of ECU200 million (US$250 million) for fusion on magnetic confinement fusion. In *particular, the report argues that Europe is neglecting the potential of an alternative approach to fusion, namely inertial confinement fusion.*

However in 1996 the EU established a Fusion Evaluation Board chaired by Barabaschi, which was commissioned as part of the preparation of the next EU 5 year budget planning cycle. The conclusions of this Board killed any hope for getting the 10% of the total fusion budget to IFE, because it recommended that the watching on ICF could be maintained at about the present level of 1-2% of the total fusion budget.

The problem of IFE in the EU arises because from its 15 members, only 5 of them are interested in the research and development of this energy source: Italy, Germany, France, United Kingdom and Spain. Portugal has joined this position lately. In these conditions it is very difficult that the UE can approve an independent programme on IFE, instead of including it in the group of *other approaches to fusion.*

To solve this unconfortable situation, it is necessary to go back to the proposal of the UNESCO, that is, to convince the scientists working in outside fields to IFE and, above all, to convince the politicians in charge of the scientific research in the different countries with strong pacifist groups, that IFE is, at least, a way as promising as the MFE in the effort to develop a new energy source abundant, safe, environmentally clean, and economical.

Specific Achievements.

The first steps in inertial confinement fusion were taken in 1966 in the Spanish Atomic Energy Commission (JEN) when I developed a radiation-transport code, named ISLERO, to analyze a ellipsoidal microcapsule with uranium walls. In its interior there was a DT micropellet with a Pu layer and, in the opposite extreme of the microcapsule, there was a hole which allowed the passing of a laser beam. To a certain extent, this microcapsule, holhraum-typed, was based on the H-bomb. Due both to the simplifications of the code Islero equations and to the inaccuracy of the parameters, the results were not reliable. For that reason I gave up temporarily this research.

Eight years later, I organized a group with a dozen of scientists to study the processes produced in the fusion of direct-driven targets based on the micropellet of DT with a layer of Pu, as I had previously done in the code ISLERO (**Neutronics of Laser Fission-Fusion Systems**, G. Velarde, JEN352, 1976, ISBN 84-500-1578-2). We developed the NORCLA code, the first non-classified coupled code, including time-dependent hydrodynamics and realistic neutron-gamma transport with adequate energy source from fusion and fission materials. Two modules composed NORCLA: NORMA (for hydrodynamics) and CLARA (for fusion-fission sources and neutron-gamma transport).

NORMA in its first version (adapted using CHART-D code from Sandia as a base) considered one-dimensional time-dependent evolution of fluids/plasmas under flexible boundary pressure conditions, supposedly formed by laser-matter interaction with the pellet. By means of a Lagrangian scheme, momentum and energy equations were solved including terms on radiation diffusion and suprathermal electrons in a simple mode. Interchange of energy terms among the different species were also included. Equation of State (EOS) and atomic coefficients were considered through analytical solutions using the average atom ionisation model (ANEOS package, improved by some specific materials using data from bibliography for high densities and temperatures). Shock waves were treated using the concept of artificial viscosity (von Neumann & Richtmyer) by adjustment up to quadratic terms. A key step to follow the desired physics of implosion was the possiblity to launch a multistep pressure profile in the boundary of the spherical target, in order to get a synchronised coalescence of all the pressure waves in the centre of the highly compressed target. This task was performed through a new additional simulation code named SINCRO. The results from NORMA (density, temperature, velocity, position) for each computational zone were treated as input for the CLARA module of burnup and transport processes.

The CLARA module includes a detailed one-dimensional and time-dependent treatment of the neutron-gamma transport equations. A key aspect of this module was the Legendre decomposition of nuclear data (cross section) in angular dependence (P_N) and the decomposition of neutron fluxes in discrete ordinates (S_N). The above characteristics helped to achieve an extremely high quality neutron transport and energy deposition of such particles in the very high compressed fusion targets and fusion-fission targets. The most advanced energy multigroup description of nuclear data was used in the code CLARA. This code included the energy and particle generation by fission reaction (when fusion-fission targets were analyzed) and the energy and particle generation by fusion reaction, after considering a simple formulation dependent on density and temperature for each zone. The final results from CLARA were the energy and power deposited in each zone of the target.

A modification of the NORCLA code was also considered by coupling the NORMA module with a MonteCarlo time-dependent system (TIMOC-ESP/LIBERTAS) in which we included burnup equations for fusion and fission.

In 1980, the JEN focused all its efforts on nuclear fusion research in the magnetic confinement field. For this reason our group decided to leave the JEN and to create the Institute of Nuclear Fusion (DENIM) at the Polytechnical University of Madrid to continue with inertial fusion energy. Since then, we have extended the research to the following areas: radiation fluidynamics (two-dimensional transport code ARWEN, using the discrete ordinates scheme in multigroups of energy, whose algorithms have been optimized for working under an AMR system); atomic physics (JIMENA and ANALOP codes, NLTE screened hydrodynamic model and detailed configuration code M3R); safety and environment (ACAB code); materials and reactor chambers; advanced fuels for IFE and tritium environmental analysis (according to the expected emission source from fusion reactors).

Acknowledgements.

I would like to thank publicly two outstanding scientists and two outstanding friends: Academician Nicolai Basov and Professor Chiyoé Yamanaka who have hardly worked along this difficult way towards an international collaboration and whose courage and effort have been a great example for all of us.

To finish, I would like also to evoke the personality and energy of Professor Edward Teller, one of the most outstanding nuclear physicists of this century. All of us are aware of his extraordinary strength, both physical and moral. Thank you very much Professor Teller for your example and friendship.

GEORGE B. ZIMMERMAN
1997

Monte Carlo Methods in ICF

George B. Zimmerman

Lawrence Livermore National Laboratory
Livermore, California 94550

Abstract. Monte Carlo methods appropriate to simulate the transport of x-rays, neutrons, ions and electrons in Inertial Confinement Fusion targets are described and analyzed. The Implicit Monte Carlo method of x-ray transport handles symmetry within indirect drive ICF hohlraums well, but can be improved 50X in efficiency by angular biasing the x-rays towards the fuel capsule. Accurate simulation of thermonuclear burn and burn diagnostics involves detailed particle source spectra, charged particle ranges, inflight reaction kinematics, corrections for bulk and thermal Doppler effects and variance reduction to obtain adequate statistics for rare events. It is found that the effects of angular Coulomb scattering must be included in models of charged particle transport through heterogeneous materials.

ANGULAR BIASING IN IMPLICIT MONTE CARLO

For decades the Implicit Monte-Carlo (IMC) method (1) has been used to simulate radiation transport in complicated multidimensional geometries. Its principal advantage over deterministic methods is the ease of implementation of all relevant physical effects, while its principal disadvantage is statistical noise.

Implicit Monte-Carlo has proved to be a valuable tool for the radiation transport in integrated hohlraum calculations of indirect drive Inertial Confinement Fusion target experiments. In these calculations laser deposition and radiation transport are solved simultaneous with the symmetry, implosion and burn of the fuel capsule, but the impact of statistical noise on the symmetric implosion of the small fuel capsule is difficult and expensive to overcome.

Here we present an angular biasing technique in which an increased number of low weight photons are directed at the imploding capsule. The method is further enhanced by directing even smaller weight photons at the polar regions of the capsule where small mass zones make the calculations most sensitive to statistical noise.

CP406, *Laser Interaction and Related Plasma Phenomena*: 13th International Conference,
edited by G. H. Miley and E. M. Campbell
© 1997 The American Institute of Physics 1-56396-696-4/97/$10.00

Biasing Toward a Sphere

One of the most important aspects of IMC, the feature that makes it implicit, is the use of effective scatters as a replacement for a fraction of the absorption and emission. This stabilizes fluctuations in the material energy and guarantees positive material temperatures, but requires the use of numerical scattering even when the physical Compton scattering is negligible. Since scattering changes a photon's direction and since we are interested in controlling the weights of photons traveling in certain directions, it is clear that photon weights must be allowed to change during the scatter. This, of course, also requires that photons be statistically created and destroyed in the scatter.

The original IMC method assumed that exactly one photon came out of each scatter. This was an advantage in terms of simplicity (sizes of vector and census stacks could be precalculated), accuracy (exact energy conservation was possible) and speed (vectorization of the scattering process was straightforward). In order to implement angular biasing it was necessary to make changes in each of these areas: The sizes of particle storage areas were allowed to grow by using modern dynamic memory management methods. Exact energy conservation was replaced with statistical energy conservation which could then be used as an accuracy check. Vectorization of the scattering process was eliminated and placed at a low priority since most of the angular biased IMC calculations were to be done on workstations without vector processing units.

The methods used to sample the angularly biased photon distribution from volume and surface emission sources are detailed below. A method to handle Compton scattering has not yet been developed, but should be amenable to reasonably efficient rejection techniques. Currently one must turn off Compton scattering when using angular biasing.

Volume Emission and Effective Scatters

In the IMC method both volume emission and effective scatters produce an isotropic angular photon distribution with a $\sigma_v B_v$ energy distribution, where σ_v is the frequency dependent absorption cross-section and B_v is the Planck function. Providing angular biasing of isotropic emission toward a spherical object, and even allowing for frequency biasing, is straightforward.

We establish a coordinate system in which the photon emission point is at the origin and the bias sphere center is on the z-axis. We take the photon importance to be $B(\mu)$, a function only of μ, the cosine of the photon direction relative to the z-axis. Sampling the photon direction consists of finding μ, where $R = \int_{-1}^{\mu} B(x)dx / \int_{-1}^{1} B(x)dx$ and R is a uniform random variable on (0,1). The weight of the resulting photon is proportional to $1/B(\mu)$. In practice we have taken $B(\mu)$ to be a two step histogram corresponding to a spherical object with a core of one importance surrounded by a halo of another importance surrounded by the universe of unit importance.

Surface Source Emission

Emission from user defined surface sources has an angular distribution that depends on the angle between the photon direction and the surface normal. We have taken this to be a general power law, $I(\vec{\Omega}) \propto (\vec{\Omega} \cdot \vec{n})^\alpha$, where $\alpha=1$ for the usual cosine surface distribution. Providing for the angular biasing of such a distribution toward a spherical object is not straightforward because, in general, the direction of surface normal and the direction to the bias sphere are not the same. We have chosen to use a rejection scheme in which angularly biased isotropic photons are created just as for volume emission, then they are rejected if $max(0, \vec{\Omega} \cdot \vec{n})^\alpha < R$. The efficiency of this rejection process is $1/(2\alpha+2)$, or 25% for the usual cosine distribution. This would be very inefficient as $\alpha \to \infty$, so surface normal emission is coded as a special case.

Biasing Toward Polar Regions

In typical two dimensional axially symmetric Lagrange hydrodynamics calculations spherical objects are represented by equal angle zoning in the r-z plane. Zones near the poles have less mass than those near the equator and thus are more subject to statistical noise problems when irradiated with equal weight photons. The obvious solution is to enhance angular biasing in a way that directs more lower weight photons at the polar regions of the bias sphere.

For K equal angle zones in 90 degrees uniform irradiation of equal weight photons results in the polar zone receiving $\pi/4K$ as many photons as the equator zone. This is 1/25 for K=20. If one were to force the same number of photons to strike each angular zone -- keeping fixed the number hitting the equator -- it would take a total of only $\pi/2$ times as many photons. The potential gain then is $8K/\pi^2$ or 16 for K=20.

Unfortunately, this is risky business. Assigning photons weights based on the latitude of intersection with a bias sphere surface will only work well if the capsule is a hard sphere with infinite opacity inside and zero opacity outside. Density gradients on the capsule surface, frequency dependent opacities and the desire to specify a bias sphere as an envelope containing several disjoint objects all mean that there is a significant probability for photons to penetrate the bias sphere. In particular, Figure 1 shows that photons entering the bias sphere near the equator can be absorbed near the polar region of a smaller sphere. This means that photon weights must be assigned based on the minimum polar angle of the trajectory within the bias sphere, not just the polar angle of the trajectory intersection with the bias sphere surface.

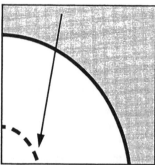

FIGURE 1. Photons directed at the equatorial region of the outer bias sphere may actually be absorbed near the polar regions of a smaller concentric sphere.

It would seem that assigning photons weights proportional to $\sin\theta_{min}$, where θ_{min} is the minimum polar angle of the trajectory within the bias sphere would automatically assure that each angular zone, whose mass is proportional to $\sin\theta$, would receive equal statistics. The problem is that the number of photons required to do this goes like $\int d\theta/\sin\theta$ and is logarithmically divergent. Also, if rejection techniques are used to establish the photon directions, then a lower limit on the photon weight must be established.

Fortunately, there is little advantage in achieving photon weight reductions within the smallest polar zone, so placing a lower limit on $\sin\theta_{min}$ of about $1/K$ should not adversely affect the statistics. We will call this lower limit $1/P_{max}$ and evaluate its optimal value in test calculations. Using such a limit the total number of photons required -- leaving the number hitting the equator unchanged -- is no longer divergent, but is larger than the non-polar biased case by a factor of about $(1 + 2/\pi \bullet \log P_{max})$. The expected gain from polar biasing then is about $P_{max}/(1 + 2/\pi \bullet \log P_{max})$ where $P_{max}\sim K$. For $P_{max}=20$ this gain is 6.9, significantly less than the 16 possible for a hard sphere capsule, but still quite substantial.

Polar biasing is implemented entirely by rejection techniques. Using the previously described methods we sample photons uniformly toward the bias sphere with importance increased by a factor of P_{max}. For each photon we then evaluate θ_{min}, the minimum polar angle of the trajectory within the bias sphere, assign an importance of $B = min(1/\sin\theta_{min}, P_{max})$ and reject the photon if $B < RP_{max}$. The efficiency of this rejection method is about $(1 + 2/\pi \bullet \log P_{max})/P_{max}$ or about 14%

for $P_{max}=20$. The importance of this efficiency depends on the amount of work that will be done with each accepted photon, but for very large P_{max}, particularly if coupled with the surface source rejection method, a search for a more efficient algorithm may be warranted.

Angular Biasing Test Problems

A series of test problems have been run to 1) prove that these angular biasing techniques do not affect any physical results, 2) confirm the reduction in statistical noise and 3) determine the optimal value for P_{max}, the upper limit on polar biasing. The geometry is a spherical annulus of vacuum with inner radius $R_0=0.35$ which is used to tally and remove any photons striking it and an outer radius of 1.0 from which a source launches photons in a cosine distribution relative to the local surface normal. The spherical annulus is represented by $K=20$ equal angular zones in 90 degrees and the distribution of source intensity on the outer radius is $P_0 + 0.1$ P_2, where P_L represents the L'th Legendre moment. One million photons were used in each simulation. The exact analytic result (2) for the intensity distribution on the inner radius is $P_0 + 0.055062 P_2$.

Table 1. Concentric sphere biasing test problems

	No biasing	Bias uniform toward sphere	Polar biasing, $P_{max}=20$
P_0	.995(.003)	.999(.001)	.997(.001)
P_2	.057(.006)	.057(.002)	.054(.003)
P_4	.009(.009)	.001(.003)	.003(.004)
P_6	.001(.010)	.003(.004)	.003(.004)
P_8	.022(.012)	.002(.004)	.002(.005)
N eq.	9555	71328	58951
N pole	444	3172	20156
$\Delta f/f$ eq.	.0102	.0038	.0053
$\Delta f/f$ pole	.0475	.0179	.0072

In Table 1 we list the Legendre moment expansion of the flux striking the inner radius with standard deviations given in parenthesis. Also given are the number of photons striking the equator and pole zones on the inner radius and the relative error in the flux at those zones. We see that all of the test problems give the analytic result of $P_0 + 0.055062\ P_2$ to within their statistical errors. Because of the cosine surface source emission the fraction of energy striking the inner radius is $1/R_0^2$, or 0.1225 for R_0=0.35. For isotropic emission it would be a factor of 4 smaller. Indeed, for the unbiased run about 12% of the one million photons did strike the inner radius, while virtually all photons struck the inner radius in the biased cases. A simple estimate of the expected standard deviations is $\sqrt{(2L+1)/N}$, where L is the moment number and N is the number of (assumed equal weight) photons striking the inner radius. This fits the data in Table 1 quite accurately, although the polar biased case has somewhat larger errors due to the use of photon weights that vary by a factor of P_{max}. Polar biasing does not help reduce the variance in the moment expansion of the flux.

Turning our attention to equator and pole zonal tallies we see that for the non-polar biased cases the number of photons striking these zones is simply related to their fractional area, $\pi/2K = 0.0785$ for the equator and $0.5(\pi/2K)^2 = 0.0031$ for the pole, and that the relative errors are precisely given by $1/\sqrt{n}$, where n is the number of photons striking the zone. For the polar biased case we see that the number of photons striking these two zones are more equal and that the error at the pole has been improved at the expense of the error at the equator. At the equator the error is larger than $1/\sqrt{n}$ because of the variation in the weights of the photons. Of course, for this test problem one could have used a polar biasing scheme that assumed a hard sphere and assigned photon weights according to the latitude of intersection with the bias sphere. This would have resulted in equal number of photons striking the equator and pole zone and the relative error in each would have been $\sqrt{K/N} = 0.0045$. Such a scheme was not implemented because it did not appear to be robust enough for real problems.

This leaves us with the problem of determining the optimal value for P_{max}. We have run a series of calculations with K=20 varying P_{max} from 1 to 100 and compared the relative polar error, $\Delta f/f$, in Table 2. Also given are the cpu times (sec) and an overall figure of merit (FOM) of $cpu \bullet (\Delta f/f)^2$. The cpu times increase with P_{max} because of the rejection algorithm used in establishing the photon directions. The relative polar error minimizes for $30 < P_{max} < 40$, but is very flat for $P_{max} > 20$ where the FOM is optimal. Although the FOM from this test problem series is not directly applicable to other problems where a different amount of work may be done with each accepted photon, the fact that optimal FOM occurs near the smallest P_{max} that achieves most of the reduction in $\Delta f/f$ is reason enough to choose P_{max}=20 as the optimal value. Although such a detailed study has not been carried out for different values of K it is anticipated that P_{max}=K is always near optimal.

Table 2. Determination of optimal P_{max} for K=20

P_{max}	$\Delta f/f_{pole}$	cpu time (sec)	FOM
1	.0179	585	.187
2	.0154	669	.159
5	.0113	738	.094
10	.0089	796	.063
15	.0076	883	.051
20	.0072	934	.048
25	.0070	1018	.050
30	.0069	1078	.051
40	.0069	1165	.055
50	.0070	1312	.064
60	.0070	1431	.070
100	.0073	1862	.099

Discussion of Angular Biasing

We have not yet discussed the optimal choice for $B(\mu)$, the importance of photons as a function of their cosine relative to the bias sphere center. In the test problem we simply made $B(\mu)$ large for all μ that would intersect the inner sphere, but in real hohlraum problems it will be necessary to use some photons to calculate the evolution of the hohlraum walls. The optimal choice is problem dependent, but it seems clear that using half of the total number of photons for the capsule and half for the hohlraum walls cannot be more than a factor of two away from the optimum. This can be accomplished by setting $B(\mu)=4\pi/\Delta\Omega$ for all μ that would intersect the capsule, where $\Delta\Omega$ is the solid angle of the capsule as seen from the wall. The overall effective gain in computing power due to angular biasing then is half this value times the gain from polar biasing, or $\frac{2\pi}{\Delta\Omega} \bullet \frac{K}{1 + 2/\pi \cdot \log K}$ if we ignore the rejection costs. For a capsule radius of 0.35 times the hohlraum radius and K=20 angular zones in 90 degrees this gain is greater than 100. We expect that 50 can be achieved in real problems.

MONTE CARLO METHODS FOR BURN PARTICLE TRANSPORT

For decades the Monte Carlo method has been used to simulate the transport of particles produced by nuclear reactions. Its principal advantage over deterministic methods is the ease of implementation of all relevant physical effects, while its principal disadvantage is statistical noise. In this section we reexamine many aspects of Monte Carlo transport and suggest several improvements to the commonly used techniques.

First the thermonuclear source of particles is presented as a Monte Carlo simulation of the reaction process. We then examine target Doppler effects from both bulk and thermal motions and give detailed formulas for charged particle energy loss rates. Finally, we demonstrate some variance reduction methods and present models for thermal and Monte Carlo transport through heterogeneous materials.

Thermonuclear Source of Particles

Particles generated by thermonuclear reactions can be handled by a straightforward Monte Carlo simulation of the reaction process, resulting in the correct correlations of the energies and angles of all outgoing particles. This is done by noting that the interaction of the two reactant Maxwell distributions can be written as a product of one Maxwellian representing the center of mass velocity, ϑ_c, using the total mass, $M = m_1 + m_2$, and another Maxwellian representing the relative velocity, ϑ_r, using the reduced mass, $m_r = (m_1 m_2)/(m_1 + m_2)$. The algorithm then consists of 1) sampling ϑ_c from its Maxwell distribution isotropically in the fluid frame, 2) sampling ϑ_r from the Gamow peak, 3) allocating the center of mass frame available energy, $E_r + Q$, to the reaction products in a momentum conserving fashion and 4) transforming from the center of mass back to the fluid frame.

Sampling from the Gamow Peak

The relative energy must be sampled from the Gamow peak distribution

$$f_{Gamow}(E_r) = \sigma(E_r)\sqrt{E_r} f_{Max}(E_r) \tag{1}$$

which is usually a narrow bell shaped function caused by a rapidly rising $\sigma(E_r)$ and a rapidly falling $f_{Max}(E_r)$. Rather than depend on having nuclear data to sample the Gamow peak we have assumed a Gaussian shape and forced its position and width to yield the correct mean, $\langle K \rangle$, and variance, $\langle K^2 \rangle$, of the relative kinetic energy:

$$\langle K^m \rangle = \frac{\int_0^\infty K^{m+1} \sigma(K) e^{-K/T} dK}{\int_0^\infty K \sigma(K) e^{-K/T} dK}. \tag{2}$$

These moments are related to derivatives of the thermonuclear reaction rates, $\langle \sigma v \rangle (T)$, by (3)

$$\langle K \rangle = T^2 \frac{d}{\partial T} ln(T^{3/2} \langle \sigma v \rangle), \tag{3}$$

$$\langle K^2 \rangle = \langle K \rangle^2 + T^2 \frac{\partial}{\partial T} \langle K \rangle. \tag{4}$$

This requires twice differentiable fits to $\langle \sigma v \rangle (T)$, but the fits do not need to be the same as those used to actually compute the thermonuclear reaction rates. In practice we have used the astrophysical form

$$\langle \sigma v \rangle (T) = f(T) T^{-2/3} e^{-(B/T)^{1/3}} \tag{5}$$

where

$$B = \frac{54\pi^4 e^4}{kh^2} Z_1^2 Z_2^2 \frac{m_1 m_2}{m_1 + m_2} \tag{6}$$

and $f(T)$ is a slowly varying function. Doing a more accurate job of sampling from the Gamow peak is not required since the dominate feature of the reaction product distribution function, a width proportional to $\sqrt{TQ}$, is obtained without any information on the Gamow peak. It is determined entirely by the center of mass frame transformation. Knowing the mean kinetic energy, $\langle K \rangle$, shifts the distribution slightly by an amount of order T/Q. The use of $\langle K^2 \rangle$ is probably not necessary, but may improve things in the wings of the distribution.

Allocating the Available Energy

If the reaction breaks up into two particles, then momentum conservation determines the energy allocation in the center of mass frame. For three or more particles we use a statistical breakup model which maximizes the entropy subject to momentum and energy constraints. This model does not use any nuclear structure information which means that it can be used for any reaction without needing additional data. On the other hand, it will not duplicate known features such as the

8.64 MeV peak in the neutrons from $t + t \to 2n\alpha$ caused by the branch that goes through the He^5 ground state.

The energy distribution of a particle of mass m from a statistical breakup of N bodies of total mass M is

$$F_N(E)dE = \sqrt{E}(E_{max} - E)^{\frac{3N-8}{2}} dE \qquad (7)$$

where the maximum possible energy, $E_{max} = (E_r + Q)(1 - m/M)$, is independent of N and occurs when all the rest of the particles go out in the opposite direction all with the same velocity required to conserve momentum. For $N = 2$, Eq. (7) is a δ-function, while for $N > 2$ the peak of the distribution, $E_{peak} = E_{max}/(3N - 7)$, is at a smaller fraction of E_{max} as N increases. As $N \to \infty$, $F_N(E)$ becomes a Maxwellian.

In order to maintain energy and angle correlations between outgoing particles we use Eq. (7) to sample only the first particle, choosing its angle isotropically in the center of mass frame. The rest of the particles go off in the opposite direction in a cluster with the kinetic energy necessary to conserve momentum and internal energy necessary to conserve energy. In the frame of this cluster the next particle is sampled isotropically from an $F_{N-1}(E)$ distribution using the cluster internal energy in place of reaction available energy. This process is continued until a two body breakup can be used to terminate the chain. Numerical experiments have shown that this method is independent of the order of choosing particles.

In practice we have found that for $N \geq 4$ it is simpler to sample the N particles from N independent isotropic Maxwell distributions at an arbitrary temperature, enforce momentum conservation by adding a velocity vector and finally scale the resulting velocities to conserve energy in the center of mass frame. We have not mathematically proved that this reproduces Eq. (7), but have used numerical experiments to show that it works.

Target Doppler Effects

Bulk Doppler Motion

One advantage of Monte Carlo is that it is easy to change frames and therefore one can simulate each part of the transport process in its most appropriate reference frame. Cross-sections must be evaluated in the fluid frame where the correct relative velocity is available, while particle storage should be done in the laboratory frame if rezoning operations are allowed. The appropriate frame in which to calculate particle motion is less clear.

Transport in the laboratory frame always results in the correct particle trajectory. After each hydrodynamics step, however, a search must be performed to find the zone containing each particle. This is not only expensive, it also makes it possible for particles to move into physically inaccessible regions, such as photons "transporting" through optically thick matter. This is not possible when transporting in the fluid frame since it requires no zonal search (unless rezoning has occurred). Fluid frame transport does, however, lead to trajectory errors if the fluid velocity varies over the path of a particle during a single time step. We have chosen fluid frame transport as the lesser of the two evils, but the best answer would be to account for mesh motion within the time step as the particle is tracked. This method is in use by one dimensional codes and, although the distance to zone boundary calculation is complicated, it may still be reasonable in higher dimensions.

Another problem associated with fluid frame transport is that discontinuous fluid velocities at zonal interfaces can make it so particles exiting one zone cannot enter its neighbor. We have eliminated this "total internal reflection" problem by breaking the mesh into triangles and interpolating between nodal velocities to form a continuous fluid velocity field. When crossing slide lines only the normal component of the fluid velocity is continuous, but that is enough to eliminate this problem.

Thermal Effects

Simulating the interaction of particles with a hot medium involves two calculations: 1) determining the correct distance to collision and 2) handling the resulting collision kinematics.

The distance to collision can only be found by integrating over all relative velocities that can be created by the background isotropic Maxwell thermal distribution. This integration yields the "hot cross-section" (4),

$$\sigma(v, T) = \frac{1}{v^2}\left(\frac{\beta}{\pi}\right)^{1/2} \int_0^\infty \sigma(v_r, 0)v_r^2 dv_r \bullet [e^{-\beta(v-v_r)^2} - e^{-\beta(v+v_r)^2}] \qquad (8)$$

where $\beta = M_{target}/(2kT)$.

This is similar to Doppler broadening of lines in forming radiative opacities. Any attempt to form a distance to collision by sampling a single representation of a background particle's energy and angle will invariably lead to an overestimate of the distance (5). We have chosen to form $\sigma(v, T)$ in the generator from cold data and then interpolate linearly in $\sqrt{T}$. This is not adequate for $T < 1ev$ since the cold data is already spread out by its group structure and since molecular effects are not included.

Once a collision has occurred we can perform the kinematics by sampling an energy and angle for the target. This is sampled from a $v_r \sigma(v_r) f_{Max}(v)$ distribution, not simply a Maxwellian, since we must favor those energy/angle pairs which correspond to v_r's with large collision rates. In practice, we sample from an isotropic Maxwellian and use rejection to favor the desired energy/angle pairs.

As an example of a thermal effects calculation Figure 2 shows the spectrum of deuterons upscattered by 14 MeV neutrons for both cold and $T = 200\text{keV}$ thermal targets. This spectrum was produced by injecting the neutrons down the axis of a very long and thin deuterium wire. This geometry is ideal for studying such collision details since each neutron interacts exactly once while all scattered particles escape freely.

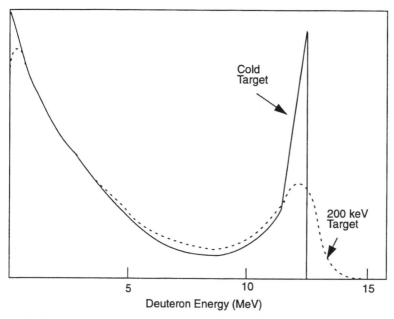

Figure 2. Deuteron upscatter from 14 MeV neutrons for cold and $T = 200\text{keV}$ thermal targets.

Charged Particle Transport Model

Charged particles transported by Monte Carlo include isotopes of H and He and user defined light and heavy ion beams. This variety of projectiles coupled with a large variety of background conditions means that the models for energy loss and scattering must be very general.

Stopping Power

Charged particle stopping power (or energy loss per unit distance) is the sum of terms due to thermal ions, free electrons and bound electrons. The electron stopping formula allows for any ratio of particle to electron velocity, includes electron degeneracy and accounts for partial ionization of both the background and projectile. In total we have

$$\frac{\partial E}{\partial x} = \frac{\partial E}{\partial x}\bigg|_I + \frac{\partial E}{\partial x}\bigg|_F + \frac{\partial E}{\partial x}\bigg|_B . \tag{9}$$

where

$$\frac{\partial E}{\partial x}\bigg|_I = \frac{4\pi e^4 Z_p^2}{V_p^2}\sum_i \frac{n_i Z_i^2}{m_i} L_i , \tag{10}$$

$$\frac{\partial E}{\partial x}\bigg|_F = \frac{4\pi e^4 Z_p^2}{m_e V_p^2} n_F L_F, \text{ and} \tag{11}$$

$$\frac{\partial E}{\partial x}\bigg|_B = \frac{4\pi e^4 Z_p^2}{m_e V_p^2}\sum_i n_i Z_i^B L_i^B \tag{12}$$

are respectively the contributions due to ions, free electrons and bound electrons. Here, Z_p is the projectile charge, V_p is its velocity and Z_i^B is the number of bound electrons on background ion i. The ion stopping is only important when $V_e \gg V_p \gg V_i$, so $L_i = ln(\lambda_D/b_i)$ is the normal ion Coulomb logarithm where

$$\lambda_D^2 = \frac{kT_e}{4\pi e^2 n_e} , \tag{13}$$

$$b_i^2 = \left(\frac{h}{4\pi m_r V_p}\right)^2 + \left(\frac{e^2 Z_i Z_p}{m_r V_p^2}\right)^2 , \text{ and} \tag{14}$$

$m_r = (m_i m_p)/(m_i + m_p)$ is the reduced mass. Stopping against free electrons, however, requires handling the full range of $y = V_p/V_e$ and results in (6)

$$L_F = \frac{1}{2}ln(1 + \Lambda_F^2)\left(\text{erf}(y) - \frac{2}{\sqrt{\pi}}ye^{-y^2}\right) , \tag{15}$$

where

$$\Lambda_F = \frac{4\pi m_e V_e^2}{\hbar \omega_{pe}} \cdot \frac{0.321 + 0.259 y^2 + 0.0707 y^4 + 0.05 y^6}{1 + 0.130 y^2 + 0.05 y^4} ,\tag{16}$$

$$\omega_{pe}^2 = \frac{4\pi e^2 n_e}{m_e}, \text{ and}\tag{17}$$

$$V_e = \sqrt{\pi} \frac{\hbar}{2\pi m_e} [4 n_e (1 + e^{-\mu/T_e})]^{1/3} , \text{ or}\tag{18}$$

$$V_e = \sqrt{(2kT_e)/m_e} , \text{ if nondegenerate.}\tag{19}$$

Assuming that the projectile velocity much exceeds the orbital velocity of bound electrons, we have (7) $L_i^B = ln((2m_e V_p^2)/\overline{I}_i)$, where the average excitation energy of bound electrons is approximated as

$$\overline{I}_i = Z_i \frac{0.024 - 0.013(Z_i^B/Z_i)}{\sqrt{Z_i^B/Z_i}} \text{ keV.}\tag{20}$$

We assume that light projectiles are fully stripped. For projectiles with nuclear charge $Z_p^{nuc} \geq 10$, however, we use the equilibrium ionization model (8)

$$Z_p = Z_p^{nuc} min \left[\frac{Z_v^2}{1.6593 - 0.1796 Z_v + Z_v^2}, 1 \right] ,\tag{21}$$

where

$$Z_v = \frac{\hbar}{2\pi e^2} (Z_p^{nuc})^{-0.69} \sqrt{V_p^2 + V_e^2} .\tag{22}$$

Ion Coulomb Collisions

In addition to energy loss, collisions with background ions also deflect the charged particles in a manner causing their perpendicular velocity to diffuse. This is implemented by deflecting the particle by an angle $\Delta\theta \propto \sqrt{\Delta x}$ after each path Δx. For this to work properly it is important to limit the path lengths so that $\Delta\theta < 90°$. These deflections effectively shorten the particle's range and are most important when simulating small scale length effects in hot, high Z materials.

The energy loss to ions also results in an upscatter source of background ions with a spectrum $N(E) \propto 1/E^2$. These upscattered particles have typical energies of a few hundred keV, so they thermalize almost instantaneously. They do, however, have a non negligible probability of causing reactions while in flight.

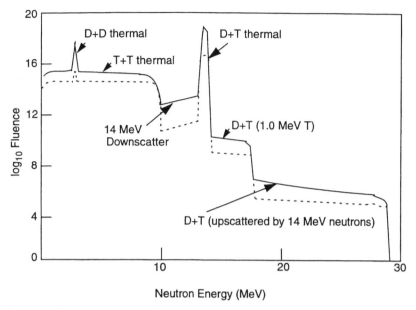

Figure 3. Neutron output and statistical uncertainty from a uniform DT sphere with $\rho r = 10^{-3}$ g/cm^2, $T = 1$ keV and $\rho = 1$ g/cc. Nearly 14 orders of magnitude in dynamic range are achieved by forcing neutron and charged particle collisions.

Variance Reduction

In order to produce meaningful results from Monte Carlo algorithms it is often necessary to employ variance reduction techniques to enhance the signal to noise. So far we have found it useful to provide weight biasing by particle type and region, force a minimum number of particles per zone, provide special exponential attenuation calculations to create pinex pictures and enhance reaction product statistics by forcing neutron and charged particle collisions. This last technique consists of multiplying a cross-section by a large number and then reducing the weights of the reaction products correspondingly. Figure 3 shows an example of its use in calculating the neutron output from a highly idealized ICF experiment. The purpose of the experiment is to measure the compression of a sphere of DT by detecting the >20 MeV neutrons produced by D's and T's upscattered by 14 MeV neutrons. Both neutron and charged particle cross-sections were enhanced by 10^4 -

10^5 to force an adequate number of collisions in this very thin system.

The uncertainty in neutron output, also shown in Figure 3, was calculated by a technique known as batching in which the output was tallied for several independent batches of the original 14 MeV neutrons. The standard deviation of the mean of these results is a measure of the uncertainty. This batching technique is also available for more realistic time dependent calculations, such as those involving hydrodynamics and other physics, but it is less clear what this measures since the hydrodynamics is not performed separately for each batch. The only correct treatment for these realistic time dependent problems is to run them several times with different random number seeds and then compute an uncertainty. When we have done this the results tend to agree with the batching method.

Transport Through Heterogeneous Materials

Small scale length heterogeneous materials may be present initially, as in the case of wetted foam ICF targets, or may develop as a result of hydrodynamic instabilities. In either case, if we can characterize the heterogeneity by a statistical model it should be possible to simulate its effects on transport processes. Here we will assume, without any real justification, the binary Markovian model, describe its effects on thermal physics and test various Monte Carlo models for simulating charged particle transport in such a mixture.

The Binary Markovian Statistical Model

The binary Markovian model (9) basically assumes that the matter can be characterized by two components, A and B, and that the probability of a straight path $\hat{x}$ staying entirely within A is

$$Q_A(\hat{x}) = e^{-|\hat{x}|/\lambda_A} \tag{23}$$

where $\lambda_A = \lambda/V_B$, V_B is the volume fraction of component B and λ is the scale length parameter of the mixture. The volume fractions are determined by assuming that components A and B are in thermal and pressure equilibrium.

The relative amounts of components A and B, the scale length parameter, λ, and the evolution of these quantities in space and time are not being considered here. We instead calculate the effects of these quantities on transport processes, providing a means of validating any models that do attempt their prediction.

Effects on Thermal Physics

Although not directly relevant to a discussion of Monte Carlo methods, the effects of a binary Markovian mixture on the thermal properties of the background medium can impact burn product transport by modifying thermonuclear reaction rates and charged particle stopping powers. Also, if we want to use these methods to validate any particular mix model, it is important to include all effects of that model.

An approximate method of performing radiation flow through a binary Markovian mixture consists of defining an effective opacity (10)

$$\sigma_{eff} = \frac{\sigma_A V_A + \sigma_B V_B + \lambda \sigma_A \sigma_B}{1 + \lambda(\sigma_A V_B + \sigma_B V_A)} . \qquad (24)$$

In the case of electron and ion conduction this same model leads to an effective conductivity

$$\kappa_{eff} = \kappa_A V_A + \kappa_B V_B , \qquad (25)$$

where we have used the large λ limit of Eq. (24) since heterogeneous material, as opposed to atomically mixed material, by definition has a scale length large compared to thermal mean free paths.

The electron-ion energy exchange rate in a heterogeneous material is simply the volume integral of the exchange rates in each component. This can be larger than the rate in an equivalent atomically mixed material by a factor $\langle n_e n_i Z_i^2 \rangle / [\langle n_e \rangle \langle n_i Z_i^2 \rangle]$ due to clumping of the electron density into the high Z component. Of course, assuming that the temperatures are independent of component while maintaining an electron-ion temperature separation may not be justified in reality.

Finally, even without the Monte Carlo charged particle transport methods described in the next section, it is possible to model charged particle loss from component A due to thermalization in component B by assuming straight line energy loss (11). If the stopping powers in the two components have the same energy dependence the probability that a particle created uniformly in A will thermalize in B is

$$L_{A \to B} = \frac{T_A}{T_A + T_B}[1 - e^{-(T_A + T_B)}] , \qquad (26)$$

where $T_X = R_X / \lambda_X$, and R_X is its range in component X.

Monte Carlo Charged Particle Transport in Heterogeneous Materials

Here we describe three statistical models of charged particle transport and compare them with an idealized test problem. The test problem is a binary Markovian mixture with $\lambda = 0.01$ cm in which a density 156 g/cc gold component B, $V_B = 0.2$, is in thermal and pressure equilibrium with a density 86 g/cc DAu$_{.01}$ component A. The temperature is held at 10 keV and the test is to calculate the correct fraction of the 1.0 MeV tritons produced by the $d + d \rightarrow p + t$ reaction in component A that come to rest in component B. The "correct" answer is approximated by running a real geometry problem in which a 0.0094 cm radius sphere of B is embedded in a 0.0161 cm radius sphere of A. This geometry has the same volume fractions and surface to volume ratio, $4 / (\lambda_A + \lambda_B)$, as the binary Markovian mixture, but it is not unique and does not get other moments of Markovian statistics correct. Nonetheless, its result of a 3.4% triton loss will be used for comparison.

The simplest Monte Carlo charged particle transport model, Model (0), uses no knowledge of the heterogeneous character of the material. It is equivalent to an atom mix model or $\lambda = 0$ and yields a triton loss of 28.4%, a factor of eight too large. This is larger than the volume fraction of component B because the electron density, and thus stopping power, is larger in B.

Model (1) uses a single bit to tell it whether the charged particle is in A or B. The bit is changed if the distance to component boundary, sampled from Eq. (23), is smaller than distances to other events. This model should yield correct results if the particles undergo straight line energy loss until they are thermalized, but results in a 8.8% triton loss, a factor of three too high, for our test problem. In order to ascertain whether the straight line energy loss is the cause of this error the real geometry problem was rerun without any angular Coulomb scattering. The resulting 10.1% triton loss is close to the 8.8% from Model (1), confirming the cause of the discrepancy. The lack of angular scattering allows tritons to travel too far from their place of birth before thermalization. Too large a fraction of them encounter component B and become thermalized in it.

In order to account for angular Coulomb scattering Model (2) was built to retain knowledge of the distances to the component boundaries in both the + and - directions along the particle path. A new event, that of randomly selecting the + or - direction every $90°$ angular scattering mean free path, was also introduced. This gave the method the ability to simulate the diffusion equation in slab geometry, but it is certainly not unique. A proper simulation of Markovian statistics would sample the entire three dimensional geometry upon the particle's birth and then carry out the transport in that geometry. Remembering the + and - distances to the edge of the current component is the minimum system with both the straight line and diffusion limits. For our test problem Model (2) yields a 3.4% triton loss exactly the same as the real geometry simulation.

Discussion of Monte Carlo Burn Transport

We have revisited many aspects of using Monte Carlo methods used to simulate particle transport in complex geometries. A method of simulating the thermonuclear reaction process was presented which maintained the correct energy/angle correlations of all outgoing particles. The effects of bulk Doppler motion are best handled by tracking the particles in the moving geometry, but the necessary distance to zone boundary calculation may be too cumbersome in two and three dimensions. Thermal Doppler effects can only be handled correctly through the use of "hot cross-sections" and even then one must perform the collision kinematics using a properly sampled target.

Detailed formulas for charged particle energy loss were given including the effects of partial ionization of background and projectile, electron degeneracy and any ratio of particle to electron velocity. Some very simple variance reduction schemes were presented, including an example in which 14 orders of magnitude in dynamic range were achieved.

Assuming a simple but realistic model for heterogeneous materials an attempt was made to calculate all of its effects on both thermal and nonthermal transport. Charged particle transport in hot, high Z heterogeneous material is particularly troublesome since the particle range, angular scattering mean free path and scale size of the mixture can all be comparable. Angular scattering was found to be an essential ingredient in calculating triton loss. Use of charged particle radiochemistry detectors would be similarly influenced.

ACKNOWLEDGMENT

This work was performed under the auspices of the U. S. Department of Energy by the Lawrence Livermore National Laboratory under Contract W-7405-ENG-48.

References

1. Fleck, J. A., Jr. and Cummings, J. D., *J. Comput. Phys.*, **8**, 313-342 (1971).
2. Haan, S. W., UCRL-ID-118152, Lawrence Livermore National Laboratory (1994).
3. Brysk, H., *Plasmas Physics*, **15**, 611 (1973).
4. Cullen, D. E. and Weisbin, C. R., *Nucl. Sci. Eng.*, **60**, 199 (1976).
5. Canfield, E. H., UCIR-694, Lawrence Livermore National Laboratory (1973).
6. Maynard, G. and Deutsch, C., *J. Physique*, **46**, 1113 (1985).
7. More, R. M., UCRL-84991, Sec. VII, Lawrence Livermore National Laboratory (1981).
8. Bailey, D. S., private communication, Lawrence Livermore National Laboratory (1987).
9. Levermore, C. D., Pomraning, G. C., Sanzo, D. L., Wong, J., *J. Math. Phys.*, **27**, 2526 (1986).
10. Levermore, C. D., Pomraning, G. C. and Wong, J., *J. Math. Phys.*, **29**, 995 (1988).
11. Levermore, C. D. and Zimmerman, G. B., *J. Math. Phys.*, **34** 4725 (1993).

DAVID SHVARTS
1999
(With co-authors)

Scaling Laws of Nonlinear Rayleigh-Taylor and Richtmyer-Meshkov Instabilities in Two and Three Dimensions

D. Shvarts[1,2], D. Oron[1,2], D. Kartoon[1,2], A. Rikanati[1,2], O. Sadot[1,2], Y. Srebro[1,2], Y. Yedvab[1,2], D. Ofer[1], A. Levin[1], E. Sarid[1], G. Ben-Dor[2], L. Erez[2], G. Erez[2], A. Yosef-Hai[2], U. Alon[3], L. Arazi[4]

[1]Nuclear Research Center Negev, Israel
[2]Ben Gurion University, Beer-Sheva, Israel.
[3]Princeton University, Princeton, NJ, USA.
[4]Tel Aviv University, Tel-Aviv, Israel.

The late-time nonlinear evolution of the Rayleigh-Taylor (RT) and Richtmyer-Meshkov (RM) instabilities for random initial perturbations is investigated using a statistical mechanics model based on single-mode and bubble-competition physics at all Atwood numbers (A) and full numerical simulations in two and three dimensions. It is shown that the RT mixing zone bubble and spike fronts evolve as $h \sim \alpha \cdot A \cdot gt^2$ with different values of α for the bubble and spike fronts. The RM mixing zone fronts evolve as $h \sim t^\theta$ with different values of θ for bubbles and spikes. Similar analysis yields a linear growth with time of the Kelvin-Helmholtz mixing zone. The dependence of the RT and RM scaling parameters on A and the dimensionality will be discussed. The 3D predictions are found to be in good agreement with recent Linear Electric Motor (LEM) experiments

1. Introduction

The Rayleigh-Taylor(RT) instability [1] occurs when a light fluid supports a denser fluid in a gravitational field, or, more generally, when a pressure gradient opposes a density gradient. The related Richtmyer-Meshkov(RM) instability [2, 3] occurs when a perturbed interface between two fluids is impulsively accelerated by a shock wave. These instabilities are of crucial importance in achieving energy gain by Inertial Confinement Fusion (ICF) [4]. Under unstable conditions, small perturbations on the interface between the fluids grow into bubbles of light fluid penetrating the heavy fluid and spikes of heavy fluids penetrating the light fluid. At very late time, due to bubble merger and the Kelvin-Helmholtz (KH) instability, which occurs when two fluids flow in proximity to each other with different velocities [5], a turbulent mixing zone (TMZ) is formed. We are primarily interested in the TMZ growth rate.

In the multimode RT instability, both the bubble and spike front heights scale with time as gt^2 , where g is the driving acceleration. The large-scale structure in the mixed region exhibits a self-similar behavior with this scale [6, 7]. In this case, it is natural that gt^2 is the only dimensional length scale of the problem, after the initial conditions are forgotten. The KH instability also evolves self-similarly according to the only dimensional length scale, $\Delta u \cdot t$ where Δu is the velocity difference between the two fluids. In contrast with the RT and KH instabilities, the impulsive nature of the RM instability does not induce such a well defined, self-similar law of TMZ evolution. In the present paper we briefly review our recent work regarding the late time scaling of the RT, RM and KH instabilities [8-13] and examine the differences induced by the dimensionality.

In systems of a single-mode periodicity the flow is governed, even at late times, by identical bubbles or eddies evolving in proximity with each other. In the RT case, bubbles growing with a constant velocity are formed, while for the RM case, bubbles growing with a velocity proportional to $1/t$ are formed [9]. For the KH case, eddies of constant asymptotic size are formed.

When a multi-mode initial perturbation is introduced, the single-mode periodicity is broken, and the flow evolves differently. An inverse cascade process arises, in which larger and larger structures are continually generated [5, 8, 10-15]. In the RT and RM instabilities, the fundamental cause of the inverse cascade is competition between structures caused by the reduced drag of large structures. In the KH instability it is caused by an effective attractive potential between neighboring vortices.

This large structure production mechanism may be viewed either as a real-space bubble/eddy competition [5, 10] or as Fourier-space mode coupling and saturation [16, 17]. In the present review we describe the evolution of the instabilities in real-space, using a statistical-mechanics merger model where the elementary particles are bubbles and spikes in the RT and RM cases and eddies for the KH case. The model is based on modeling the front by an array of particles (bubbles and spikes or eddies), each evolving according to it's single-particle asymptotic behavior obtained from Layzer's potential flow model [18] for the RT and RM instabilities at A=1, and from a vortex model for the low Atwood RM [19, 20] and KH instabilities. In the model, particles overtake smaller neighboring ones to form larger particles ("particle merger") [10] at a rate calculated using extensions of the potential flow model [9] or the vortex model [13] describing two-particle competition. The model predicts self-similar asymptotic growth of the TMZ, resulting in a scaling law. The particle size distribution, normalized to the average particle size, reaches a fixed distribution. In the case of the RT instability, the 2D multimode bubble front was found to grow asymptotically as $h_{2D} = \alpha_{2D} \cdot A \cdot g \cdot t^2$ with $\alpha_{2D} \cong 0.05$ in accordance with many known experimental and numerical results [6, 7, 10, 21]. The RM bubble front was found to grow according to the power law: $h_b = a_0 \cdot t^{\theta_B}$ with $\theta_{B,2D} = 0.4$ for all values of A, where a_0 depends on the initial perturbation [10], in agreement with 2D numerical simulations and preliminary experimental results [21]. In the KH case, the TMZ grows as $h = c \cdot \Delta u \cdot t$, with $c = 0.2$ at low Atwood numbers [5].

The model predicts similar growth in 2D and 3D for the RT instability, but different scaling parameters for the 3D RM mixing zone growth, with values of $\theta_{B,3D}$ appreciably lower than the 2D values. These results are in agreement with more recent Linear Electric Motor (LEM) experimental results [22, 23].

In section 2 the two-dimensional statistical mechanics merger model and its results for the RT, RM and KH instability evolution at all Atwood ratios will be discussed. Section 3 deals with the differences between the evolution of 2D and 3D perturbations.

2. A Two Dimensional Statistical-Mechanics Merger Model for Multiple Bubbles/Eddies Ensemble

Both the RT and the RM mixing zones are topped by column-shaped bubbles of light fluid, rising and competing. At late nonlinear stages, large bubbles grow faster then smaller ones. Therefore, a bubble adjacent to smaller bubbles expands sideways and accelerates while its neighbors shrink and decelerate until they are washed downstream.

This "bubble-merger" process leads to constant growth of the surviving bubbles and to a constant decrease in the number of surviving bubbles. This description of the mixing front was pioneered by Sharp and Wheeler [14]. Using a similar statistical model, a potential flow model for the A=1 RT and RM instabilities [9] and a vortex model for the low Atwood RM and KH instabilities [13], we present a merger model, which allows realistic merger rates and treatment of more general flow problems.

In the model, particles are arranged along a line, and are characterized by their size λ_i (bubble diameter for the RT and RM instabilities and eddy diameter for the KH case). For the RT and RM case, each bubble rises with a velocity $u(\lambda_i)$ equal to the asymptotic velocity of a periodic array of bubbles with diameter λ_i. In the KH case each eddy is assumed to have it's asymptotic size $h(\lambda_i)$. The particle competition is included using a merger rule: two adjacent particles of sizes λ_i and λ_{i+1} merge at a rate $\omega(\lambda_i, \lambda_{i+1})$, forming a new particle of size $\lambda_i + \lambda_{i+1}$. Thus, the surviving particle expands to fill the space previously taken by both particles.

For the RT and RM instabilities the mean front height is found using the average bubble velocity $dh(t)/dt = \langle u \rangle$, while for the KH case it is found using the average eddy height $h(t) = \langle h(\lambda_i) \rangle$. All averages are taken over the size distribution function $g(\lambda)d\lambda$, defined as the number of particles in the mixing zone front with sizes within $d\lambda$ of λ. Neglecting near-neighbor correlations, we can write a mean-field evolution equation for the size distribution:

$$N(t)\frac{dg(\lambda,t)}{dt} = -2g \int\limits_0^\infty g(\lambda',t)\omega(\lambda',\lambda)d\lambda' + \int\limits_0^\lambda g(\lambda-\lambda',t)g(\lambda',t)\omega(\lambda-\lambda',\lambda)d\lambda' \quad (1)$$

where $N(t) = \int\limits_0^\infty g(\lambda,t)d\lambda$ is the total number of particles at time t. The first term on the RHS of equation 1 is the rate of elimination of particles of size λ from the front by mergers with other particles, and the second term is the rate of creation of bubbles of size λ by the merger of two smaller particles.

The model is solved using the single particle behavior and the two particle merger rate from [9] and [13], as shown in the following sections. An example of the agreement between the model and the simulation is shown on *figure 1* for the case of an ablatively driven RT bubble front.

2.1. High Atwood RT and RM instabilities

We model the instability growth from an initial random short-wavelength noise by a set of many short-wavelength modes, creating bubbles that rise and compete. Therefore, the physics of the single bubble and the interaction between two neighboring bubbles must be introduced, and serve as input to the statistical model. For high Atwood numbers (very large density ratios) we use an extension of Layzer's potential flow model [18]. In the model, the bubble evolution is assumed to be dependent only on the shape of the bubble in proximity to the tip. The incompressible flow equations are expanded to second order in the longitudinal space coordinate near the bubble tip, and solved, yielding the single bubble behavior and the two bubble merger rate [9]. For the RT instability, a single bubble initially grows exponentially and then saturates to a constant asymptotic velocity of $u = \sqrt{g\lambda/(6\pi)}$. The RM bubble grows initially with a constant

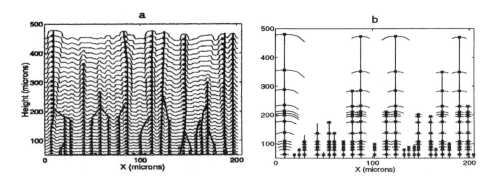

Figure 1. a) Simulation bubble front (contour of maximal light fluid penetration) evolution. Rising bubbles are denoted by '*', merger processes by lines b) Model bubble front evolution.

velocity at the linear stage, and asymptotically with a velocity of $u = 1/3\pi \cdot \lambda/t,$. Both results are in good agreement with numerical simulations and experiments.

Based on the potential flow model results, the statistical mechanics model of the previous section is analyzed. The merger model [8, 11] predicts that both the RM and RT fronts dynamics flow to a self-similar stage, where a fixed bubble size distribution (normalized to the average bubble size) is reached. At this stage, the number of bubbles and the average bubble height evolve according to:

$$\frac{dN(t)}{dt} = -\langle\omega\rangle N(t), \qquad \frac{dh(t)}{dt} = \langle u\rangle \tag{2}$$

Integrating equation 2 it is found [8, 11] that for the RT instability $h_B = \alpha_B g t^2$, where $\alpha_B = \langle u\rangle\langle\omega\rangle/(4g)$. For the RM instability a power law of $h_B = a_B t^{\theta_B}$ with $\theta_B = \langle\omega\rangle$ and a_B a constant representing the initial conditions is obtained. Applying the potential flow model results, we found that for the 2D RT instability, $\alpha_B \cong 0.05$ and for the RM instability $\theta_B \cong 0.4$.

Some of the properties of the RT and RM mixing zone can be derived from a simple drag-buoyancy equation:

$$(\rho_1 + C_a \cdot \rho_2) \cdot V \cdot \frac{du}{dt} = (\rho_2 - \rho_1)V \cdot g - \rho_2 \cdot S \cdot u^2 \tag{3}$$

where ρ_1 and ρ_2 are the two fluid densities, V the volume, S the surface area, the second term on the LHS is for added mass, and the two terms on the RHS are for buoyancy and kinematic drag, respectively. Using $V/S = \lambda/C_d$, we obtain equations for the bubble and spike front evolution. For the A=1 case, equation 3 reduces to [18]:

$$C_a \cdot \frac{du}{dt} = g - \frac{C_d}{\lambda} \cdot u^2 \tag{4}$$

The values of C_a and C_d can be derived from the potential flow model for A=1, yielding $\alpha_{B,S}$, $\theta_{B,S}$ and the average value of $h_b/\langle\lambda\rangle$ for the multimode case. Using the 2D values $C_d = 6\pi$ and $C_a = 2$, and assuming $\alpha_B = 0.05$ we obtain for the scale-invariant regime:

$$\theta_B = 0.4, \quad \frac{h_B}{\langle\lambda\rangle} = 0.27 \tag{5}$$

2.2. Low Atwood RM instability

For the RM $A \to 0$ (two fluids having similar densities) single-bubble evolution and two-bubble interaction we apply a vortex model, based on an extension of single-mode vortex models [19, 20].

In the case of Atwood number $A \to 0$, an initial cosine perturbation transforms, early in time, into a localized vortex array. This enables us to model the problem using an infinite vortex line with alternating directions [19]. The model predicts that the single mode, low A, RM asymptotic velocity is $V_{asy} = 1/2\pi \cdot \lambda/t$ [13], in agreement with [10]. Note that this is larger than the result at A=1 where, as derived earlier, $V_{asy} = 1/3\pi \cdot \lambda/t$. The difference is due to the higher inertia of the bubble in the low A case. These results are in good agreement with full 2D simulations [13], and were recently confirmed experimentally [19, 24].

The single-bubble model was extended to the case of two-bubble interaction by setting an array of four periodic infinite vortex lines creating an array of alternating large and small bubbles. The interface evolution is derived following the movement of these vortex lines, where each moves in the complex velocity field induced by the other three. From the model we can infer the merger rate $\omega(\lambda_1, \lambda_2)$.

Using the analytical results of the previous section: $\theta_B = \langle \omega \rangle$, a power law of $\theta_B = 0.4$ was found for the A=0 case, in agreement with 2D numerical results.

The results for both the RT and the RM instability are in good agreement with full 2D numerical simulations [10]. It was found that $\theta_B = 0.4$ and $\alpha_B = 0.05$ for all Atwood numbers. For the spikes, however, α_S goes from 0.5 at A=1, indicating an expected free fall behavior of the spikes, to 0.05 for A=0, where the bubbles and spikes are symmetric. Similarly, θ_S goes from 1 at A=1, to 0.4 for A=0. Assuming the dominant spikes in the spike front evolve according to the periodicity of the dominant bubbles in the bubble front, we get for intermediate values of A [10]:

$$\alpha_S/\alpha_B \approx 1 + A, \quad \theta_S/\theta_B \approx 1 + A \tag{6}$$

2.3. The Low A Kelvin-Helmholtz Instability

As mentioned earlier, the KH instability occurs when two fluids flow in proximity to each other with different velocities (a shear layer flow). Experiments performed in the 1970's by Brown and Roshko [5], and more recently by others [25], in planar geometry, demonstrated that the TMZ grows linearly with time according to $h(t) = c(\rho_1/\rho_2) \cdot \Delta u \cdot t$, where ρ_1 and ρ_2 are the fluid densities, Δu the velocity difference between the fluids and t the time, and that vortex pairing is the main mechanism of eddy enlargement [15].

Therefore, it is evident that a statistical merger model, similar to the one applied for the RM and RT instabilities, can be used to model the KH instability. Looking at the KH instability at the zero average velocity reference frame, it reduces into a formation of a set of vortices. Due to the shear flow, all vortices are rotating in a given direction and only differ in strength.

The single-mode initial perturbation KH instability is represented by an infinite line of equal strength equally spaced vortices. A similar analysis was performed for two eddy merger, obtaining the merger rate as a function of the vorticity ratio between the two eddies.

The statistical model predicts an asymptotic self similar eddy size distribution, which is in very good agreement with experimental data [5, 25], and linear growth for the TMZ evolution. The linear growth coefficient of the whole TMZ was found to be 0.2, in very good agreement with the experimental value.

3. Three dimensional effects

Recent LEM experiments [22, 23] have confirmed the predicted scaling laws for the RT case but resulted in somewhat different scaling parameters for the RM case, mainly $\theta_B \approx 0.25$, substantially lower than the value of 0.4 predicted by the statistical model. Also, the scale-invariant parameter $h_b/\langle\lambda\rangle$ obtained in these experiments was higher by a factor of 3-4 than the predicted value.

These differences can be attributed to three-dimensional effects. The change in the surface to volume ratio between the 2D case and the 3D case results in different values for the parameters C_d and C_a in equation 4 for the A=1 case, from which one obtains a different relation between α_B, θ_B and $h_B/\langle\lambda\rangle$. Assuming, $\alpha_B \cong 0.05$, a result obtained from 3D experiments [6, 22, 23, 26] and simulations [27, 28], we get in the 3D case (using $C_d = 2\pi$ and $C_a = 1$):

$$\theta_{B,3D} = 0.2, \quad (h_B/\langle\lambda\rangle)_{3D} = 0.8 \approx 3 \cdot (h_B/\langle\lambda\rangle)_{2D} \tag{7}$$

in good agreement with the recent experimental values [22, 23]. The 3D values of α_B, θ_B and $h_B/\langle\lambda\rangle$ were all found, as in the 2D case, to depend weakly on A. The 3D values of α_S/α_B and θ_S/θ_B are also similar to the 2D value of $1 + A$, in good agreement with the experimental results [23].

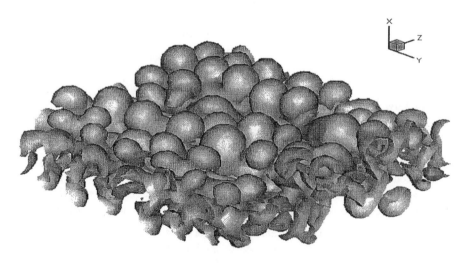

Figure 2. A typical frame from a 3D calculation.

Preliminary full 3D simulations of the RT instability indicate that the discrepancy between the experiments and the previously predicted results indeed arises from the

3D effects. *Figure 2* shows a typical frame from such a 3D simulation. Shown is a shade plot of the interface between the two fluids (A=0.5). These calculations are analyzed using Voronoi cell diagrams of the bubble tips, differentiating between the rising (accelerating) bubbles and those washed downstream (decelerating). Three consecutive Voronoi diagrams from the 3D simulation are shown in *figure 3*. Note the continuous growth of the surviving bubbles, and the fact that almost all bubbles have 4-6 near neighbors. *Figure 4a* shows the bubble and spike front heights, defined as the mean height of the leading bubbles and spikes, indicating a value of $\alpha_B = 0.046$ and $\alpha_S = 0.076$, similar to the results predicted by the 2D model. However, a comparison of the value of $h_B/\langle\lambda\rangle$ between a 2D simulation and a 3D simulation, plotted in *figure 4b* shows that the 3D value is nearly 3 times higher than the 2D one, in accordance with the experimental results.

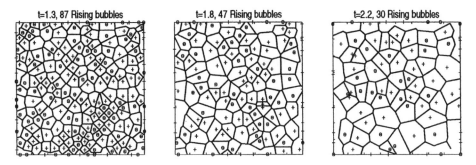

Figure 3. Three consecutive Voronoi plots of the 3D simulation evolution. Accelerating are denoted by 'O', decelerating ones by '+'.

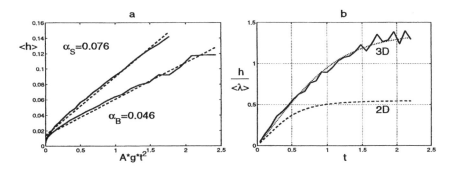

Figure 4. a) values of α_B and α_S as obtained from the 3D simulation (solid line - simulation height, dashed line - $\alpha \cdot gt^2$). b) comparison of $h_b/\langle\lambda\rangle$ between a 2D simulation (dashed) and a 3D one (full line). Dotted line is a smoothed 3D line.

More insight to the 3D effects can be gained using the bubble competition approach. There are three main effects of the dimensionality on the process. The first is the increase by a factor of 1.5-2 in the asymptotic bubble velocity and the resulting decrease of the merger time due to the shortening of time scales. The second is due to the geometrical difference between 2D and 3D: since the conserved quantity when two bubbles merge is area rather than length, the merger process becomes $\lambda_1, \lambda_2 \to \sqrt{\lambda_1^2 + \lambda_2^2}$ in 3D rather than $\lambda_1, \lambda_2 \to \lambda_1 + \lambda_2$ as in 2D. The third effect is related to the number of neighbors of a single bubble. in 3D this number is 4-6 (rather than 2 in 2D), as can be seen from the Voronoi diagrams of *figure 3*. The last two effects result in narrowing of the bubble size distribution and thus reduce $\langle w \rangle$, since it takes bubbles of similar sizes longer time to merge. All three effects have a small effect on the value of α in the RT case, but lower the expected value of θ_B substantially in the RM case.

4. conclusion

Using a statistical merger model of large coherent structures, each evolving by itself, and competing with it's neighbors, we obtain the scaling laws for the TMZ evolution of the RT, RM and KH instabilities. The RT and KH mixing zones are characterized by a single scale length, gt^2 and $\Delta u \cdot t$, while the RM mixing zone is characterized by two scale lengths, t^{θ_B} for the bubble front, and t^{θ_S} for the spike front.

Three dimensional effects do not change the RT scaling parameter α_B appreciably, but lower the RM parameter from $\theta_{B,2D} = 0.4$ to about $\theta_{B,3D} = 0.2$. The bubble to spike asymmetry is $\alpha_S/\alpha_B \approx \theta_S/\theta_B \approx 1 + A$ in both 2D and 3D. The 3D value of the scale-invariant parameter $h_B/\langle\lambda\rangle$ was found to be larger by a factor of 3 than the 2D value, indicating growth of much more elongated bubbles.

References

[1] Lord Rayleigh, Proc. London Math. Soc. 14, 170 (1883); Scientific Papers (Cambridge University Press, Cambridge, 1900), Vol. II,p. 200.
[2] R. D. Richtmyer, Commun. Pure. Appl. Math. 13, 297 (1960)
[3] E. E. Meshkov, Fluid Dyn. 4, 101 (1969).
[4] S. W. Haan, Phys. Plasmas 2, 2480 (1995).
[5] G. L. Brown, A. Roshko, J. Fluid Mech. 64, 775 (1974).
[6] D. L. Youngs, Physica D 12, 32 (1984).
[7] N. Freed, D. Ofer, D. Shvarts, S.A. Orszag, Phys. Fluids A 3, 912 (1991)
[8] U. Alon, J. Hecht, D. Mukamel, D. Shvarts, Phys. Rev. Lett. 72, 2867 (1994).
[9] J. Hecht, U. Alon, D. Shvarts, Phys. Fluids 6, 4019 (1994).
[10] U. Alon, J. Hecht, D. Ofer, D. Shvarts, Phys. Rev. Lett. 73, 534 (1995).
[11] U. Alon, D. Shvarts and D. Mukamel, Phys. Rev. E 48, 1008 (1993).
[12] D. Oron, U. Alon, D. Shvarts, Phys. Plasmas 5 (5), 1467 (1998).
[13] A. Rikanati, U. Alon, D. Shvarts, Phys. Rev. E 58, 7410 (1998).
[14] D. H. Sharp, Physica D 12D, 3 (1984); J. Glimm, D. H. Sharp, Phys. Rev. Lett. 64, 2173 (1990).
[15] L. P. Barnel, Phys. Fluids 31, 2533 (1988).
[16] S. W. Haan, Phys. Rev. A 39, 5812 (1989); S. W. Haan, Phys. Fluids B 3, 2349 (1991).
[17] D. Shvarts, U. Alon, D. Ofer, R.L. McCrory, C.P. Verdon, Phys. Plasmas 2, 2465 (1995).
[18] D. Layzer, Astrophys. J. 122, 1 (1955).
[19] J. W. Jacobs, J. M. Sheeley, Phys. Fluids 8, 405 (1996).
[20] N. Zabusky, J. Ray, R.S. Samtaney, Proceedings of the 5th Int'l Workshop on Compressible Turbulent Mixing, R. Young, J. Glimm, B. Boston editors, World Scientific, p. 89 (1995).
[21] G. Dimonte, M. B. Schneider, Phys. Rev. E 54, 3740 (1996).
[22] M. B. Schneider, G. Dimonte, B. Remington, Phys. Rev. Lett. 80 (16), 3507 (1998).
[23] G. Dimonte, Phys. Plas. 6, 2009 (1999).
[24] O. Sadot, L. Erez, U. Alon, D. Oron, L.A. Levin, G. Erez, G. Ben-Dor, D. Shvarts, Phys. Rev. Lett. 80 (8), 1645 (1998).
[25] A.R. Paterson, "A First Course in Fluid Dynamics", Cambridge University Press (1983); J.W. Naughton, L.N. Cattafesta and G.S. Settles, J. Fluid Mech. 330, 271 (1997).
[26] K. I. Read, Physica D 12, 45 (1984).
[27] D. L. Youngs, Laser and Particle Beams 12, 725 (1994).
[28] J. Hecht, D. Ofer, U. Alon, D. Shvarts, S. A. Orszag, R. L. McCrory, Laser and Particle Beams 13, 423 (1995).

STEVEN W. HAAN
1999
(With co-authors)

Design of ignition targets for the National Ignition Facility

Steven W. Haan, T. R. Dittrich, M. M. Marinak, and D. E. Hinkel
Lawrence Livermore National Laboratory
Livermore CA 94550

Abstract

This is a brief update on the work being done to design ignition targets for the National Ignition Facility. Updates are presented on three areas of current activity : improvements in modeling, work on a variety of targets spanning the parameter space of possible ignition targets ; and the setting of specifications for target fabrication and diagnostics. Highlights of recent activity include : a simulation of the Rayleigh-Taylor instability growth on in imploding capsule, done in 3D on a 72° by 72° wedge, with enough zones to resolve modes out to 100 ; and designs of targets at 250eV and 350eV, as well as the baseline 300 eV ; and variation of the central DT gas density, which influences both the Rayleigh-Taylor growth and the smoothness of the DT ice layer.

Introduction

The National Ignition Facility is a large laser being built at Lawrence Livermore National Laboratory. One of its principal goals is to demonstrate the ignition and burn of targets containing deuterium-tritium fuel. Work has been going on for about 10 years on the design of these targets ; this article is an update on the status of that work. Results are presented in four areas : improvements in the modeling of the baseline targets ; design work at different drive temperatures ; setting of target fabrication specifications ; and simulations being used to define the diagnostic instruments needed for ignition implosions.

The basic ideas behind the ignition targets are described in refs. [1-4]. The laser is described in other articles in this proceedings. The baseline ignition target is shown *in figure 1*, with radiation drive and laser power shown in *figure 2*.

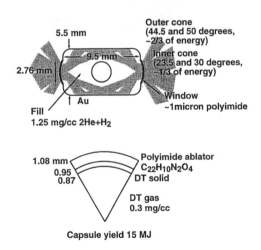

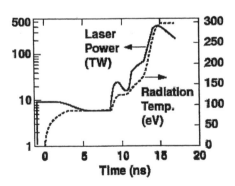

Figure 1. Baseline indirect drive ignition target. This target uses 1.3 MJ of absorbed laser light.

Figure 2. Laser power and radiation temperature driving the target shown in *figure 1*.

Improvements in modeling

The design simulations for the NIF targets have mostly been done with two codes, HYDRA [5] and LASNEX [6]. LASNEX is a 2-dimensional radiation hydrodynamics code with very complete modeling of most relevant physical processes. HYDRA is a relatively new 3D radiation hydrodynamics code with modern hydrodynamics algorithms and Arbitrary Lagrange Eulerian capability. It has all of the physics necessary for NIF ignition capsule implosions, including multi-group radiation diffusion, electron conduction, alpha particle transport, and neutron energy deposition. HYDRA has been used quite extensively in analysis of Nova experiments and has been well verified. [5,7]

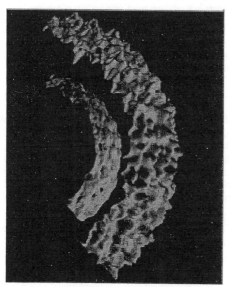 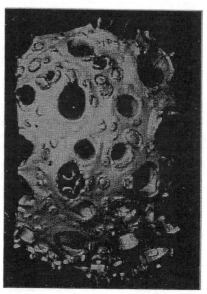

Figure 3. Density isocontour at 13.6 g/cc of imploding NIF shell prior to ignition, in HYDRA simulation by M. Marinak. The ice and ablator surface were roughened with 1 μm and 20 nm perturbations respectively, with spectra based on characterization of real surfaces.

Figure 4. Same simulation during burn (1MJ cumulative yield at this time). The total yield for this simulation is 15.4 MJ

The recent simulations have been done on a 72° by 72° wedge, with enough zones to resolve modes up to 100. Simulations were done on a beryllium-ablator target, the same target used in simulations reported previously.[8] Initial perturbations were similar to those used in other work. As described in ref. 6, they are based on characterization of Nova capsule surfaces, and of DT ice layers in _beta-layering experiments. Initial roughnesses were 1 μm rms for the ice, and 20 nm rms for the beryllium, with a multi-mode spectrum initializing modes from around 2 (1/2 wave in 72°) to 100. The capsule performs well in the simulation, producing 15.4 MJ of fusion yield. *Figures 3* and *4* show density isocontours at a time prior to ignition and after burn is under way, respectively. *Figure 3* shows the growth of short wavelength perturbations to amplitudes somewhat less than the shell thickness. *Figure 4* shows two interesting features. There are low density regions where bubbles of hot fuel have risen through the decelerating shell. There are also crater-like features on the outer surface. At these points, fingers of beryllium have reached down to the stagnation shock, which defines the outer isocontour. Outside of the craters the density isocontour is in DT ; inside the craters, it is in beryllium, with the material interface forming the crater. Both of these features are also seen in 2D multimode simulations, although the nonlinear perturbation growth is quantitatively larger in three dimensions. Also, of course, the features are 2D rings in the 2D simulations, rather than being round spots as in 3D. This work will be described in more detail in a future publication by M. Marinak, who both wrote the code and did the simulations.

Design work at the edges of parameter space : 250eV and 350eV

Most NIF target design work has been at 300 eV, which is the temperature currently considered most likely to produce ignition. However, the optimum tradeoff of the various physics issues may result in our operating at some other temperature, and it is important for us to understand the issues and requirements at the different temperatures. By intent these designs stress some physics issues and will not look robust in all regards.

A design at 350eV, by D. Hinkel, is close to being a 0.65 scale of the baseline. The hohlraum is 6.6 mm long by 3.56 mm diameter. The capsule is beryllium, outer radius 710 μm, with a layer of clean Be and a layer of 1% Cu-doped Be This target performs well in integrated simulations, giving 70% of clean 1D yield. The integrated simulations required increasing case-to-capsule ratio from 2.5 to 3 (at waist) in order to avoid both hohlraum filling and asymmetric pressure on capsule from the hohlraum fill gas. At case-to-capsule ratio 2.5, various gas fills were considered; at low gas fills, the hohlraum filled unacceptably, while at higher gas fills a pressure spike on axis perturbed the capsule causing an asymmetric implosion even with the radiation on the capsule artificially symmetrized. designs operating at 300 eV or lower there is an operating regime between these two failure modes, but at 350 eV the case-to-capsule ratio has to be increased before a satisfactory gas-fill can be found. Estimates of filamentation in the plasma typical of this hohlraum, done with the code pF3D, suggest we may need SSD with as much as 6 Å of bandwidth (at 1 μm) to control the filamentation. Polarization smoothing is also being considered, and may reduce the bandwidth requirement. Rayleigh-Taylor analysis of this target shows very low growth. It can tolerate 5 micron DT rms and >50 nm ablator rms.

A very small design at 250eV has been described previously [9] ; somewhat bigger designs, corresponding to 1.3 MJ of laser light, have now been considered in more detail. As was the case for the design described in ref. 7, the Rayleigh-Taylor growth is quite severe. An important option at 250eV may be to increase the capsule coupling efficiency, as described by Suter elsewhere in this proceedings. It may be possible to increase the energy absorbed by the capsule to as much as four times what was assumed in the 250eV work to date. This will make the Rayleigh-Taylor growth acceptably small, although detailed design work on this remains to be done. Suter's proposed reoptimization of the hohlraum may also allow for quite high gains, close to 100 MJ, from 250eV targets.

Target Fabrication Specifications

In addition to Rayleigh-Taylor analysis for each target we are considering, some work has been done dedicated to determining target fabrication specifications in two areas : low mode deviations from sphericity, and the performance vs. central gas density.

The low-mode specification is based on linear analysis. We did a set of linear-regime single mode simulations with various modes in order to determine a growth factor for each mode number. We also did large-amplitude simulations with the various modes to determine how large a final amplitude could be tolerated for each mode. These results can be combined, with a safety factor, to determine a table of effective maximum amplitudes for each mode. There remain details to work out before it would be appropriate to publish the table, so it will not be presented here.

The other area of work to describe here is considerations of the central gas density. It appears likely, from work on characterization of the DT ice layer, that the quality of the surface is a strong function of the operating temperature. The surface is likely to be smoother at temperatures close to the triple point, at 19.8 K at 0.62 mg/cc, while the target performance has a somewhat weaker contrary dependence. Previously the designs have usually stipulated a

central density of 0.3 mg/cc, corresponding to an operating temperature of 18.3 K. The Rayleigh-Taylor sensitivity of the baseline 300 eV polyimide design is shown in *figure 5*, contrasting the growth for the two different gas densities. While the performance suffers somewhat at higher gas fill, and the ablator surface specification is somewhat tighter, it appears that there is an option of operating at the triple-point density of 0.6 mg/cc. This may be necessary if the surface roughness is more than 3 μm at 18.7 K, it having been characterized at about 1 μm rms at 19.8K.

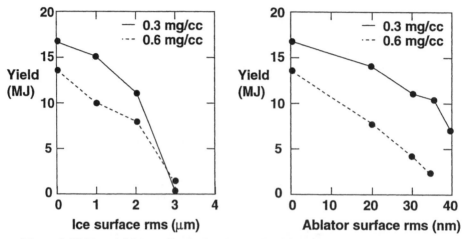

Figure 5. Yield vs. initial amplitude showing the Rayleigh-Taylor instability sensitivity for the 300 eV polyimide capsule, at two different operating temperatures (equivalent to two different central gas densities, as indicated). The higher density of 0.6 mg/cc is close to the triple-point gas density, while most design work has been done at 0.3 mg/cc, corresponding to a temperature 1.3 K lower. Although performance is worse, and specification tighter, at the higher gas density, the ice surface may be significantly smoother at this temperature and the optimum operating point may be the higher temperature and density.

Diagnostic specifications

A final area of active work is the simulation of target output as needed to optimize the diagnostic instruments being planned for the ignition implosions. Proposed diagnostics include :

➤ the usual neutron yield, with time-of-flight broadening to give ion temperature ;

➤ secondary and tertiary neutrons to get ρr measurements (the details of this are an important topic for near-term future work) ;

➤ the charged particle output (results from Petrasso et al. on Omega suggest that this will be very interesting, and they have proposed high energy protons as a shell ρr measurement [10] ;

➤ a reaction history measurement, possibly based on the direct γ-rays produced at a small rate in D-T reactions ;

➤ imaging in neutrons and high energy x-rays, as discussed below.

We would like to put together a detailed description of how each of these would manifest various possible failure modes, so that we can say ahead of time how any likely failure mechanism will be diagnosed. That will help us to set specifications for the instruments

High resolution images would be very valuable, both in neutrons and x-rays. It is likely that the size of the images will be valuable ; for example, shock timing problems will probably result in images that are bigger than expected, while mix will probably reduce the size of the images. If they are nonreproducibly out of round, that will suggest problems with power balance, low mode target fabrication problems, or too-large laser-plasma instabilities, which would have exactly the same effect on the implosion. If the images are reproducibly out-of-round, it will suggest problems with intrinsic hohlraum asymmetry or « aligned » target fabrication issues such as the cryo layer roundness or the tent supporting the capsule.

Besides this rather large list of possible inferences from each case, work on images has also suggested that the inference to be made is not necessarily simple. For example, an implosion with a P2 asymmetry in target fabrication (the shell too thin at the poles) comes in, bounces on the poles, and produces an image that is somewhat elongated along the poles (« sausaged ») and pinched in around the waist. So images will have to be interpreted carefully, probably with code simulations as guides.

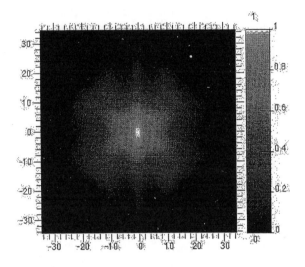

Figure 6. Simulated neutron image, brightness vs. position in microns.

One possible image, typical of what we might want to see, is in *figure.6*. This is a simulation that was made to « fizzle » by running without alpha particle deposition. Similar images have been made in simulations with too-large P_6 and P_8 asymmetries. The simulation shown also includes short-wavelength surface perturbations in even modes 2 through 40. It produced 27 kJ of yield (9.7×10^{15} neutrons), with peak ion temperature 9.4 keV and burn-weighted ion temperature (such as would be seen from neutron time-of-flight broadening) 4.8 keV. The structure, which is caused primarily by the P_6 and P_8 hohlraum asymmetry, is small enough that imaging with resolution of 5 µm or better would be valuable.

High energy x-ray images are quite similar to the neutron image shown. X-rays are produced by bremsstrahlung emission from the very hot fuel, and at energies above about 10 keV are transmitted through the high density fuel around the emitting hot-spot. At energies high enough to be transmitted with absorption, for example 20 keV, the image is virtually indistinguishable from the neutron image. At 8 keV, significant absorption features begin to appear. Since those tend to be in spikes that are plunging down into the hot-spot, they correspond to the less-emitting areas of the image and tend to exaggerate the structure. It would be useful to have images at a variety of energies, to reconstruct both the emission and absorption.

Summary

We are using increasingly detailed modeling, on a variety of NIF ignition targets, to understand the available parameter space and to set specifications. HYDRA 3D simulations have been done with more than adequate solid angle and resolution to ensure accurate modeling of the Rayleigh-Taylor instability growth. We have a suite of designs spanning 250eV to 350eV, with the optimum still looking like it is close to 300 eV. Finally, specifications for target fabrication and diagnostic instrumentation are challenging but appear viable.

[1]   S.W. Haan, S.M. Pollaine, J.D. Lindl, L.J. Suter, R.L. Berger, L.V. Powers, W.E. Alley, P.A. Amendt, J.A. Futterman, W.K. Levedahl, M.D. Rosen, D.P. Rowley, R.A. Sacks, A.I. Shestakov, G.L. Strobel, M.Tabak, S.V. Weber, G.B. Zimmerman, W.J. Krauser, D.C. Wilson, S.V. Coggeshall, D.B. Harris, N.M. Hoffman, and B.H. Wilde., *Phys. Plasmas* **2**, 2480 (1995)

[2]   W.J. Krauser, N.M. Hoffman, D.C. Wilson, B.H. Wilde, W.S. Varnum, D.B. Harris, F.J. Swenson, P.A. Bradley, S.W. Haan, S.M. Pollaine, A.S. Wan, J.C. Moreno, and P.A. Amendt, *Phys. Plasmas* **3**, 2084 (1996).

[3]   D. C. Wilson, P.A. Bradley, N.M. Hoffman, F.J. Swenson, D.P. Smitherman, R.E. Chrien, R.W. Margevicius, D.J. Thoma, L.R. Foreman, J.K. Hoffer, S.R. Goldman, S.E. Caldwell, T.R. Dittrich, S.W. Haan, M.M. Marinak, S.M. Pollaine, and J.J. Sanchez., *Phys. Plasmas* **5**, 1953 (1998).

[4]   J. D. Lindl, *Inertial Confinement Fusion : The Quest for Ignition and Energy Gain Using Indirect Drive*, Springer Verlag, New York (1998).

[5]   M. M. Marinak, R. E. Tipton, O. L. Landen, T. J. Murphy, P. Amendt, S. W. Haan, S. P. Hatchett, C. J. Keane, R. McEachern, and R. Wallace, *Phys. Plasmas* **3**, 2070 (1996).

[6]   G. B. Zimmerman and W. L. Kruer, *Comments Plas. Phys.* **2**, 51 (1975).

[7]   M.M. Marinak, S.G. Glendinning, R.J. Wallace, B.A. Remington, K.S. Budil, S.W. Haan, R.E. Tipton, and J.D. Kilkenny, *Phys. Rev. Lett* **80.**, 4426 (1998).

[8]   M.M. Marinak, S.W. Haan, T.R. Dittrich, R.E. Tipton, and G.B. Zimmerman, *Phys. Plasmas* **5**, 1125 (1998).

[9]   T. R. Dittrich, S. W. Haan, M. M. Marinak, S. M. Pollaine, and R. McEachern, *Phys. Plasmas* **5**, 3708 (1998).

[10]   R. D. Petrasso, C. K. Li, M. D. Cable, S. M. Pollaine, S. W. Haan, T. P. Bernat, J. D. Kilkenny, S. Cremer, J. P. Knauer, C. P. Verdon, R. L. Kremens, *Phys. Rev. Lett.* **77**, 2718 (1996).

STEFANO ATZENI
2001

A Survey of Studies on
Ignition and Burn of Inertially Confined Fuels

Stefano Atzeni

Dipartimento di Energetica, Università di Roma "La Sapienza" and INFM
Via A. Scarpa, 14, 00161 Roma, Italy

A survey of studies on ignition and burn of inertial fusion fuels is presented. Potentials and issues of different approaches to ignition (central ignition, fast ignition, volume ignition) are addressed by means of simple models and numerical simulations. Both equimolar DT and T-lean mixtures are considered. Crucial issues concerning hot spot formation (implosion symmetry for central ignition; igniting pulse parameters for fast ignition) are briefly discussed. Recent results concerning the scaling of the ignition energy with the implosion velocity and constrained gain curves are also summarized.

1. Introduction

The achievement of the target energy gain required for IFE, i.e. for energy production by inertially confined fuels, requires strong fuel compression and minimization of the energy required to initiate self-sustaining fusion reactions [1, 2]. This second condition can be obtained by means of ignition from a hot spot heated to 5–10 keV, followed by burn propagation to the whole fuel. In the standard approach to IFE [1, 2], the hot-spot is generated by hydrodynamic cumulation at the centre of the imploded fuel. In the so-called fast ignitor [3], the hot spot is created by direct heating of a portion of a precompressed fuel. A few authors [4, 5] have also proposed to rely on volume ignition of an optically thick homogeneous fuel. An important issue in IFE research is just the selection of the approach to ignition. While detailed target design relies on highly sophisticated numerical simulations, analytical models and relatively simple simulations of idealized model-problems are of great value. They provide insight, allow for understanding trends and issue, and for orientative parameter choice.

This paper summarizes studies by the author and his collaborators comparing performance and issues of the above approaches to ignition. While the fuel of most immediate concern is the equimolar DT mixture, we also briefly consider T-lean mixtures. We shall see that, with a few notable exceptions, most simulation results can be predicted semi-quantitatively by very simple models developed about two decades ago. We also discuss two topics related to the creation of the hot spot, namely, the sensitivity of central ignition to implosion asymmetries, and the beam parameters required to create the hot spot in fast ignitors. Finally, the illustration of recent results on the scaling of the ignition energy of centrally ignited fuels will show that simple models still provide answers to open problems.

2. Ignition and burn of precompressed fuel assemblies

Most aspects of the physics of ignition and burn of an ICF target can be analyzed by studying the evolution of a preassembled spherical fuel configuration with simple radial profiles of density and temperature, initially at rest. This model schematizes the fuel conditions around ignition time, and removes the difficulties associated to the study of the compression stage. The performance of the fuel assembly is then measured by the *fuel gain* $G_F = E_{fus}/E_F$, i.e. the ratio of the fusion energy to the fuel internal energy at ignition. With standard assumptions [2] on the efficiency of the driver and on the overall coupling efficiency η of the driver energy to the fuel, IFE requires $G_F \geq 1000$.

Volume igniting fuels are studied by assuming uniform initial profiles of density and temperature. Hot-spot ignition is instead modeled by spherical systems with a central hot sphere (with temperature $T > 5$ keV) immersed in a colder spherical shell. The approach in which the hot spot is created hydrodynamically at implosion stagnation is modeled by an initially isobaric configuration [6] (with the hot spot surrounded by much denser fuel, so that the pressure is everywhere uniform), while fast ignition by an isochoric configuration (with uniform fuel density).

Analytical models for the fuel gain were developed in the late 1970's and in the early 1980's by Kidder [7] and Bodner [8] for initially isochoric fuels, and by Meyer-ter-Vehn for initially isobaric fuels [9]. Rosen and Lindl [10] developed a more general model which includes the previous ones as particular cases. We shall refer to it as the MtV-RL model. It was apparent that isochoric configurations allow for larger gain (at given fuel energy) than isobaric ones, but no scheme was available at the time to exploit such a higher potential. The proposal of the fast ignitor by Tabak *et al.* [3] and studies on volume ignition by Basko [5] renewed the interest in the comparison of the potentials of the different configurations. In particular, the study discussed in the following subsections was prompted by discussions I had at I.L.E. with S. Nakai and H. Takabe in the Fall of 1993. I felt that assumptions of the models had to be reviewed and model results had to be checked by simulations.

2.1 Ignition conditions

One of the inputs to the MtV-RL model is the ignition condition (defining the parameters of the hot spot), which in general should depend on the considered configuration. Ignition conditions have been determined by means of large sets of numerical simulations [11]. They are shown in Fig. 1 in the form of ignition curves in the plane $\rho R - T$, where ρ, T, and R, are respectively, the initial density, temperature and radius of the hot region. While the curve for isochoric conditions is universal (i.e., independent on the density), the curve for isobaric conditions depends slightly on the ratio of the densities of hot and cold fuel, while the curve for volume ignition depends on the fuel mass. Figure 1 should be read as follows: if the initial conditions lies above the ignition curve, the hot spot temperature will eventually increase, its size will grow, and burn will propagate. For volume ignition, where the hot region coincides with the whole fuel, burn will occur without propagation. The ignition curves of Fig. 1 are well approximated by a simple 0-D model (see [12]; also [2, 11] and refs. therein). It consists of two equations, for the energy and mass balance of the hot spot, respectively. The first one accounts for heating by fusion products and cooling by radiation, electron conduction and mechanical work. The second equation describes cold fuel ablation by charged fusion products escaping the hot spot and by electron conduction from the hot spot. Neutron heating and partial radia-

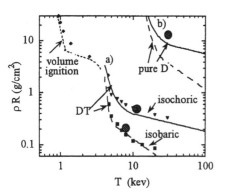

Fig. 1   Ignition conditions for DT and D fuels.

tion reabsorption are also easily introduced when appropriate. Concerning mechanical work, notice that it is initially zero in isobaric systems, while can be important in isochoric systems [11], where the hot spot has much larger pressure than the cold fuel, and hence expands rapidly. Indeed, Fig. 1 shows that isochoric ignition requires a larger hot spot than isobaric ignition. In both cases, as ρR becomes larger than about 1 g/cm^2 the ignition temperature approaches the ideal ignition temperature of 4.3 keV. For very large ρR the assembly becomes optically thick, and ignition occurs well below 4 keV. However, low temperature ignition is slow, and fuel disassembly prevents ignition of assemblies with $T < 1$ keV [13, 14]. The filled circles in the figure indicate optimal ignition points for isobaric and isochoric initial conditions. For isobaric conditions we have $\rho R = 0.2$ g/cm^2, $T = 8$ keV; for isochoric conditions

$$\rho R = 0.5 \text{ g/cm}^2 \quad T = 12 \text{ keV}. \tag{1}$$

The figure also shows the ignition thresholds for pure-deuterium. Since the hot spot energy for ignition scales as $C_v(\rho R)^3 T/\rho^2$, where C_v is the fuel specific heat, we immediately see that ignition of pure-D requires 10^4 times more energy than that of DT (at fixed ρ).

2.2 Scaling of the fuel energy gain

Gain curves for hot-spot ignited fuels are computed by inserting the above ignition conditions in the MtV-RL model. Despite the popularity of the model, it had not been checked versus simulations prior to 1993. By studying the behaviour of the gain as the pressure is varied, keeping fuel mass m and isentrope parameter α constant, we found [11] that the model is very rough at low pressure (i.e. when the total ρR is small), but is surprisingly accurate in the region of highest gain. In particular, the limiting gain curves, i.e. the curves of maximum gain for a given fuel mass and α, were reproduced very accurately by the model. We have

$$G_{F-\text{lim}}^{\text{isobaric}} \simeq 6000(\hat{E}_F/\alpha^3)^{0.3} \tag{2}$$
$$G_{F-\text{lim}}^{\text{isochoric}} \simeq 19000(\hat{E}_F/\alpha^3)^{0.4} \tag{3}$$

where the $\hat{E}_F$ is the fuel energy in MJ.

Concerning volume ignition, one cannot develop simple models because both the ignition condition and the burn efficiency depend non-trivially on the initial parameters. However, a limiting gain

$$G_{F-\text{lim}}^{\text{vol.ign.}} \simeq 1000 \hat{E}_F^{0.16} \tag{4}$$

was determined by a large set of numerical simulations [11].

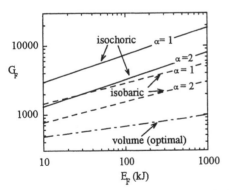

Fig. 2 Limiting gain curves for DT fuels.

From Fig. 2 we see that volume ignition is at most of marginal interest to IFE, since it can achieve $G_F = 1000$ only for fuel energy about 1 MJ. We also see that isochoric ignition confirms its potential for higher gain than isobaric ignition, although the advantage is smaller than estimated in previous works, which assumed the same ignition conditions for isobaric and isochoric conditions. The advantage may appear small (a mere factor between 2 and 3) if one compares the gain at given fuel energy, but becomes substantial (a factor of 7–10) when comparing the energy required to achieve a given gain.

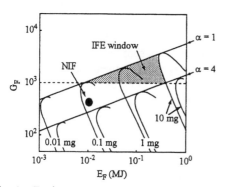

Fig. 3   Fuel energy gain vs fuel energy for two
values of the isentrope parameter α and several
fuel masses.

The wealth of information provided by
the model is apparent from Fig. 3, referring
to the isobaric case, i.e. to standard central
ignition. It shows fuel gain curves for assemblies with different masses and two values of α, as well as the relevant limiting gain
curves. We see that even without introducing
any additional constraint coming from symmetry or stability, we can identify an *IFE
window*, where the fuel gain exceeds 1000
and the mass is below 10 mg (to allow for
containment of the fusion yield). We see that
the fuel mass should be 1–10 mg, the fuel
energy between 50 and 500 kJ, and $\alpha \leq 2$.
Corresponding densities are $\rho \geq 200$ g/cm^3,
pressure $p \geq 200$ Gbar, fuel specific energy
of the order of 50 kJ/mg, implosion velocity v_i about 300 km/s (increasing as the mass decreases; the relation between E_F and v_i will be discussed in Sec. 5). The figure also shows a point
representing the parameters of the ignition experiment foreseen on the NIF [2]. The region
of lowest masses ($m < 0.1$ mg) is difficult to access, since correspond to higher the specific
energy and hence to higher implosion velocity. An upper limit to the latter is set by implosion
instability [2]. Notice, however, that in this region the gain is inadequate to IFE.

3. Energy concentration in the hot spot: issues in central and fast ignition

Creation of the hot spot requires concentrating energy in time and space and raises different
issues in central ignition and in fast ignition. In central ignition the main difficulties are related
to the symmetry and hydrodynamic instability of the implosion, while in fast ignition to the
beam intensity required to heat a small spot in a very short time. We now deal with two of the
many aspects involved.

3.1 Sensitivity of centrally ignited targets to drive nonuniformities

The previous results on ignition and energy gain refers to perfectly spherical systems. In
practice, drive and target asymmetries and the development of the Rayleigh-Taylor instability
(RTI) lead to the formation of an asymmetrical hot spot. Here we restrict attention to perturbations with long wavelength (i.e. with spherical mode number $l \leq 8$), such as those associated
to macroscopic asymmetries in the capsule irradiation pattern. Since, due to the convergence of
the fuel shell, the hot spot radius is a small fraction, typically $1/C = 1/30$–$1/40$ of the initial
shell radius, small relative pressure non uniformities result in large deformations of the hot spot.
We expect that nonuniformities of the driving pressure p_a have to satisfy a relation of the form
$\delta p_a/p_a < (1/2)(1/C) \approx 1\%$. Somewhat more insight is obtained from the following argument.
DT hot spot ignition is the result of a competition between alpha-particle heating and cooling
mechanisms. If we keep the volume of the hot spot fixed, loss terms increase as the hot spot
area increases, i.e. as the amplitude and mode number of the surface perturbation increase. The
effect of these losses on ignition depends, however, on the specific capsule. If the capsule is designed in such a way that the hot spot formed in a 1-D spherical implosion has just the minimum
parameters required for ignition, any small deviation will lead to failure; if, instead, a relative-

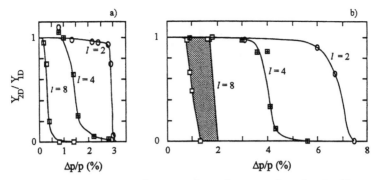

Fig. 4   Yield degradation of a reactor size target due to long wavelength 2-D effects. The figures shows the energy release computed from 2-D simulations, normalized to that computed in the equivalent spherically symmetric case, versus the peak-to-valley perturbation of the pressure driving the implosion, for modes l =2, 4, and 8. The left frame (a) refers to a marginally igniting target; the right frame (b) to the best case considered. The grey area represents an error bar mainly caused by numerical diffusion affecting the code mesh-rezoning technique.

ly large hot spot is generated, the target will be more tolerant to asymmetries. This has been confirmed by our 2D study [15, 16], which extended previous work by Verdon and McCrory [17]. However, as shown in Fig. 4, even relatively robust target cannot tolerate $l = 8$ pressure perturbations with peak-to-valley relative amplitude larger than 1% of the unperturbed value. More recent studies have shown that perturbations with $l = 12$–20 are about as dangerous as those with $l = 8$. Higher modes of stationary pressure perturbation need not be considered in the final stage of the implosion. Actually, short wavelength perturbations have large growth due to Rayleigh-Taylor instability at the ablation front, but play a minor role during stagnation. This is because of nonlinear saturation effects and because – due to the smallness of the wavelength compared to the shell thickness towards the end of the implosion – *feedthrough* of the perturbation from the outside of the shell to the inside is modest compared to long wavelength modes.

3.2 Beam parameters for fast ignition

Fast ignition relaxes requirements on symmetry and stability, but requires an ultraintense beam to create the hot spot in a short enough time. The energy E_p, power P, and intensity I to be delivered to the fuel can be roughly estimated by using the ignition condition given by Eq. (1) and taking the hot spot mass as $m_h \approx (4\pi/3)\rho R^3$, the pulse duration $t \approx E_p/P$, as $t \approx R/c_s$, (c_s: sound speed) and the beam radius equal to the hot spot radius. One thus gets $t \approx 40/\hat{\rho}$ ps, and energy $E_p \geq 72/\hat{\rho}^2$ kJ, where $\hat{\rho} = \rho/(100 \text{ g/cm}^3)$. For a better evaluation, we have performed a large parametric study based on 2-D numerical simulations, in which a hot spot is created by fast particles hitting the surface of a DT sphere precompressed to density ρ. Pulse energy is minimized by using heating particles with penetration depths in the range $0.3 \leq \mathcal{R} \leq 1.2$ g/cm^2. There is no advantage in using particles with shorter range, while the required energy grows as $\mathcal{R}$ exceeds 1.2 g/cm^2. The corresponding delivered beam parameter windows for ignition are shown in Fig. 5, for different values of ρ. For $50 \leq \rho \leq 3000$ g/cm^3 the minimum parameters for ignition are fitted by [18]

$$E_p = 140 \, \hat{\rho}^{-1.85} \text{ kJ} , \qquad (5)$$

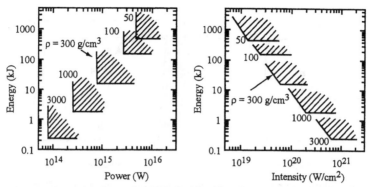

Fig. 5   Beam parameter windows for DT fast ignition in the deposited power - deposited energy plane and in the deposited intensity - deposited energy plane, fot different values of the DT density ρ.

$$P = 2.6 \times 10^{15} \, \hat{\rho}^{-1} \text{ W} , \tag{6}$$

$$I = 2.4 \times 10^{19} \, \hat{\rho}^{0.95} \text{ W/cm}^2 . \tag{7}$$

Notice that E_p given by Eq. (5) is a factor two larger at $\rho = 300$ g/cm^3 than estimated by the simple model. This is in small part (10–20%) due to 2-D effects and in major part to energy losses during hot spot heating (see Sec. III.D of Ref. [18]).

These results have recently been used to interpret 2-D studies (See [19] and Atzeni, Temporal, Honrubia, submitted) analyzing the recently proposed scheme [20] of fast ignition by laser accelerated proton beams.

Burn performance was also studied by 2D simulations. We found that the limiting gain is in very good agreement with a straightforward extension of the MtV-RL model, which reads

$$G_{\text{Ffi}} \simeq 18000(\hat{E}_F/\alpha^3)^{7/18} E_F^{0.018} = 4200 m_{\text{mg}}^{0.3}/\alpha^{0.87} \tag{8}$$

very similar to Eq. (3), which confirms the potentials of fast ignitors.

4. An option for Tritium-lean fuels

While the DT mixture is the easiest to ignite and more reactive fuel, it requires Tritium breeding and releases most of its energy in the form of fast neutrons, which activate reactor structures. Alternate fuels therefore raise a certain interest, but the prospects about their use in IFE are at best uncertain, due to high ignition temperature, lower specific yield and lower reactivity. The fast ignitor concept, with its promise for higher gain, motivated the study of non-DT fast ignitors [21]. The most appealing case, in our opinion, concerns a bulk-deuterium fuel, with a small DT seed, ignited by a pulse satisfying Eqs. (5)–(7). Detailed 2-D calculations have been performed, with accurate treatment of thermal and non-thermal reactions, of neutron transport and of the diffusion of all charged fusion products and bulk ions knocked-on by neutrons. Of course, it is found that the fuel gain decreases as the total tritium content decreases. However, an interesting result is that targets with about 20 mg of deuterium compressed to density $\rho \approx 1000$ g/cm^3 (which requires $E_F = 1$ MJ) and with total tritium fraction about 1%, achieve $G_F \simeq 1000$ and have tritium breeding ratio larger than one. This means that a reactor employing such targets would not have the complex and costly T-breeding blanket.

5. Scaling of the ignition energy with the implosion velocity

The main goal of the present stage of ICF research is the attainment of ignition, using central ignition [2]. It is well known that the minimum required driver energy E_{ig} depends significantly on hydrodynamic stability. The latter, in turn, can be related to the final value of the shell implosion velocity v_i [2]. It is therefore useful to express E_{ig} as a function of v_i. According to the isobaric model, one has $E_{ig} \propto \alpha^3 v_i^{-10} \eta^{-1}$. Notice that here the isentrope parameter refers (as in Sec. 2) to the stagnating fuel. Detailed numerical simulations of igniting capsules by Herrmann *et al.* [22], are instead fitted by $E_{ig} \propto p_a^{-0.77\pm0.03} v_i^{-5.89\pm0.12} \alpha_{if}^{1.88\pm0.05}$ [22], which shows a weaker dependence on the velocity. In addition, it expresses E_{ig} in terms of the value of the imploding shell isentrope parameter α_{ig} and of the ablation pressure p_a driving the implosion. Both these quantities can be controlled by choosing the driver parameters appropriately. The authors of Ref. [22] pointed out that, in general $\alpha > \alpha_{if}$, since the fuel entropy increases as the shell stagnates. This effect had been studied by Meyer-ter-Vehn and Schalk [23], who developed self-similar solutions for the stagnation of an imploding shell, but had not taken into account in the simple gain models.

This observation led us to introduce the results of Ref. [23] into the isobaric model [24]. First, we observe that for geometrically similar compressed assemblies, i.e. compressed fuels with the same ratio of the cold fuel radius to the hot spot radius, we have $E_{ig} \propto p^{-2}$, where p is the pressure of the stagnating fuel. The self-similar solutions of Refs. [23] and [25] show that the stagnation pressure is determined uniquely by the driving pressure and the Mach number $\mathcal{M}$ of the implosion, being $p \approx 3.6 p_a \mathcal{M}^3$, where $\mathcal{M} = v_i/c_0$, with c_0 the sound velocity in the shell. If we approximate $c_0 = \sqrt{p_0/\rho_0}$; $p_0 \simeq p_a$, and $p_0 \propto \alpha_{if} \rho_0^{5/3}$, we obtain

$$p \propto p_a^{0.4} v_i^3 \alpha_{if}^{-0.9}, \tag{9}$$

and then

$$E_{ig} \propto p^{-2} \propto p_a^{-0.8} v_i^{-6} \alpha_{if}^{1.8}, \tag{10}$$

which nearly exactly reproduces the result of Ref. [22]. This is surprising, in view of the oversimplified nature of our model. We wonder whether it is just a lucky coincidence or instead it has a deeper meaning (e.g., it could indicate that physical processes neglected in the self-similar solution, such as thermal conduction, do not affect the stagnation pressure). This point certainly deserves further studies.

Considering again Eq. (9), we notice that in ICF targets α_{if} is set by the pulse shape, while p_a is limited by instabilities in laser-plasma interaction. Therefore, for a given target concept and fixed constraints on the beam pulse, p is uniquely determined by v_i. This means that gain curves at fixed stagnation pressure are equivalent to curves at fixed implosion velocity. This observation help us to understand an old and well known, but so far unexplained result. In 1982 Meyer-ter-Vehn [9] showed that gain curves generated by the isobaric model assuming constant p and α reproduce the gain curves generated in 1979 by the LLNL laboratory by means of numerical simulations. This is of interest still today, since gain curves predicting indirect-drive target performance are very similar to the 1979 curves [2]. We now know that the LLNL curves assume fixed stability constraints, i.e. fixed implosion velocity [2]. The previous discussion tells us that these curves are just equivalent to curves drawn by taking the stagnation pressure constant. (A few words of caution are however needed: first, the parameters assumed by Ref. [9] are different from those typical of an indirect-drive target, but one can draw very similar curves by using more appropriate parameters; second, there is still a small difference since the curves of Ref. [9] assume constant α rather than constant α_{if}.)

6. Conclusion

In this paper we have surveyed our results concerning ignition and burn of IFE fuels. The surprising aspect is that simple models not only help us to understand established results, but still provide new ones and may sometimes guide present research. Of course, quantitative target design relies on sophisticated numerical simulations, in turn validated by a number of experiments performed at major laboratories throughout the world (as seen from many contributions to this volume). We hope that the experiments planned for the end of the decade on Megajoule class lasers will resolve the still open issues and eventually demonstrate ignition. In conclusion, I wish to stress that progress in this area is the result of contributions by a large number of researchers, whose work cannot be properly acknowledged, due to space limitations.

Dedication and Acknowledgements

This is the text of the lecture delivered on the occasion of the presentation of the Edward Teller medal. The work acknowledged by the award would not have been possible without the continuous support and encouragement by my family. I therefore dedicate this lecture to my wife Beatrice, my daughters Chiara and Marta, my mother Caterina and, especially, to the memory of my father, Livio. Space limitation do not allow to mention the great number of colleagues which I should thank for advice, collaboration and discussions. However, I would like to thank explicitly M.L. Ciampi, S. Graziadei, A. Guerrieri and M. Temporal, who contributed to portions of the work described here and to the development of the DUED code, which I use routinely for the 2D simulation of inertial fusion problems.

References

[1] J. Nuckolls, *et al.*, Nature **239**, 139 (1972).
[2] J.D. Lindl, Phys. Plasmas **2**, 3933 (1995).
[3] M. Tabak, *et al.*, Phys. Plasmas **1**, 1626 (1994).
[4] H. Hora and P.S. Ray, Z. Naturforsch. Teil A **33**, 980 (1978).
[5] Basko, Nucl. Fusion **30**, 1443 (1990).
[6] S. Yu. Guskov, O. Krokhin and V.B. Rozanov, Nucl. Fusion **16**, 957 (1976).
[7] R. Kidder, Nuclear Fusion **16**, 406 (1975).
[8] S. Bodner, J. Fusion Energy **1**, 221 (1981).
[9] J. Meyer-ter-Vehn, Nucl. Fusion **22**, 561 (1982).
[10] M. Rosen and J.D. Lindl, in Laser Program Annual Report 83, UCRL 50021-38, p.3-5.
[11] S. Atzeni, Jpn. J. Appl. Phys. **34**, 1980 (1995).
[12] S. Atzeni and A. Caruso, Nuovo Cimento **80B**, 71 (1984).
[13] A. Caruso, Plasma Phys. **16**, 683 (1974).
[14] G.S. Fraley, *et al.*, Phys. Fluids **17**, 174 (1974).
[15] S. Atzeni (1990) Europhys. Lett. **11**, 639 (1990).
[16] S. Atzeni (1992) Part. Accel. **37-38**, 495 (1992).
[17] R.L. Mc Crory and C.P. Verdon, in Inertial Confinement Fusion (A. Caruso and E. Sindoni, eds.), Compositori-SIF, Bologna (1989), p. 83.
[18] S. Atzeni, Phys. Plasmas **6**, 3316 (1999).
[19] S. Atzeni, *et al.*, Target studies for fast ignition by laser-accelerated proton beams, this volume.
[20] M. Roth, *et al.*, Phys. Rev. Lett. **86**, 436 (2001).
[21] S. Atzeni and M.L. Ciampi, Nucl. Fusion **37** 1665 (1997).
[22] M. Herrmann, M. Tabak and J.D. Lindl, Nucl. Fusion **41**, 99 (2001).
[23] J. Meyer-ter-Vehn and C. Schalk, Z. Naturforschung **37A**, 955 (1982).
[24] S. Atzeni and J. Meyer-ter-Vehn, Nucl. Fusion **41**, 465 (2001).
[25] A. Kemp, J. Meyer-ter-Vehn, S. Atzeni, Phys. Rev. Lett. **86**, 3336 (2001).

Teller Medal Lecture IFSA2001:

Problems and solutions in the design and analysis of early laser driven high energy density and ICF target physics experiments

Mordecai D. Rosen

Lawrence Livermore National Laboratory
University of California
Livermore, Ca 94551

The high energy density (HED) and inertial confinement fusion (ICF) physics community relies on increasingly sophisticated high power laser driven experiments to advance the field. We review early work in the design and analysis of such experiments, and discuss the problems encountered. By finding solutions to those problems we put the field on firmer ground, allowing the community to develop it to the exciting stage it is in today. Specific examples include: drive and preheat in complex hohlraum geometries with the complicating effects of sample motion; and issues in the successful design of laboratory soft x-ray lasers and in the invention of methods to reduce the required optical laser driver energy by several orders of magnitude.

1. Introduction

The field of inertial confinement fusion (ICF) naturally lives in the parameter space of high energy density (HED). Even the initial shock pressure driving a high gain capsule [1] is of order 1 megabar (1 MB) or 10^{12} erg/cm^3, followed by successive shocks nearing 100 MB that ultimately accelerates the frozen DT pusher to "thermonuclear speeds" of order 3 10^7 cm/sec. When the DT pusher then stagnates upon implosion, that kinetic energy is converted to thermal energy and fusion ensues, with local pressures exceeding 1 TB! In other applications of high power optical lasers such as drivers of exploding foil laboratory soft x-ray lasers [2], dense hot plasmas with internal pressures of some 10 MB serve as the lasing medium.

The field of HED in general, then, relies on increasingly sophisticated high power laser driven experiments in order to advance. In this paper we review early and heretofore unpublished work in the design and analysis of such experiments, and discuss the problems encountered and how solutions were found. In Section 2 we discuss x-ray drive and preheat in complex hohlraum geometries, and the complicating effects of sample motion. Section 3 will review some issues in the successful design of laboratory soft x-ray lasers and in the invention of methods to reduce the required optical laser driver energy by several orders of magnitude.

Common to both these examples was the necessity to step outside the narrow confines of the problem and to look at the whole system from a larger, fresh point of view in order to find solutions to the problems. As we shall see, in the case of the complex hohlraums, this literally required "thinking outside of the box". By finding those solutions we put the field of HED studies on firmer ground, allowing the community to develop it to the exciting stage it is in today.

2. Half-hohlraum HED physics experiments

2.1 Experimental Results

The history of early ICF implosion campaigns is briefly reviewed in Ref [1]. In the mid to late 70's much effort was expended on the 2-beam Argus laser at LLNL to drive glass encased DT spherical capsules indirectly with laser produced x-rays inside a gold cylindrical can or hohlraum. The 2 beams entered on either side of the can through laser entrance holes (LEHs) and impinged on on-axis gold cones that were meant to scatter the light in an axisymmetric pattern onto the walls of the can, in the "primary" region, where they would produce x-rays. The capsule sat in the center of the "secondary", between the 2 cones, and radiation flowed to it by passing through the annular region between the on axis scattering cone and the cylinder walls. This was a way to preserve axi-symmetry for driving the capsule (under the constraint of having only 2 beams) while not illuminating the capsule directly. The goal of the implosions was to reach the milestone of densities of some 20 gm/cm^3 or 100 X liquid density The campaign was called "Cairn" which means milestone.

As the Argus laser was 1.06 μm light, we now know in retrospect how challenging a goal "100 X" was, since hot electrons were readily (though somewhat erratically) produced by laser plasma instabilities. These hot electrons caused preheat of the capsule making high-density implosions quite difficult to achieve. However, the inside of a tiny implosion is difficult to diagnose, and doubly so when that implosion is inside a closed can. In order to study the Cairn failure mechanisms in detail, we designed experiments that effectively cut these hohlraums exactly in half, at the mid-plane, and closed off most of the open end with a back plate made of Au wall material. All that was visible then to the outside world, through a 200 μm diameter diagnostic hole in the center of the back plate, was the inner surface of the capsule hemisphere. In fact we used a flat slab of glass of equivalent thickness to the capsule pusher instead of a hemisphere. An example of these "half-Cairn" targets is shown in Fig. (1).

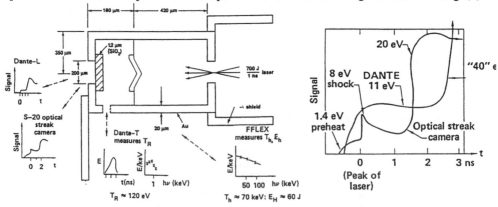

Fig. (1): Experimental set-up and drive side results.   Fig. (2): Cold side data.

The incident laser beam provided about 700 J in 1 ns to the "half-hohlraum" ("half", of course only in the sense of what the ultimate goal of the study was - understanding implosions in a "full", 2 sided illumination hohlraum. The "half-Cairn" is certainly as much a hohlraum, or x-ray oven, as a full Cairn.). On the "drive" side of the glass slab (namely the side facing the inside of the hohlraum), and local to its surface (namely in the "secondary" part of the half hohlraum, behind the gold scattering cone) we measured the radiation temperature (T_r) to be about 120 eV. We used a sub-keV 10 channel, broadband, 250 psec resolution x-ray detector named Dante that measures the emission coming from a hole in the side of the hohlraum. (The "primary" region of the half-hohlraum had a T_r [measured in a similar fashion] of about 140 eV.) Since the Dante was not spatially resolved, the can had large shields on it so that the hole is the only likely x-ray source that Dante can see (the shields block, for example, a plume of x-ray emitting plasma flowing out the LEH). We also measured the production of hot electrons by monitoring the high-energy bremsstrahlung x-rays they produced when they stopped in material, with a time and space integrated x-ray detector called FFLEX. We inferred that about 60 J of hot electrons were produced, with a temperature of about 70 keV.

On the cold or "burn-thru" side of the glass slab, namely the side facing the outside world, "peeking out" through the 200 μm diameter hole in the back plate of the half-hohlraum, we measured the temperature vs. time in 2 ways. We used another Dante spectrometer, and we also used a spatially resolving, absolutely calibrated, 15 ps resolution optical streak camera called the streaked optical pyrometer (or "stroptometer" for short). Fig. (2) shows the results of those 2 measurements. Initially these results raised even more questions than they answered! The early time signals are probably consistent, given the poorer time resolution of the Dante. The stroptometer shows a prompt preheat signal of about 1.4 eV by the laser peak. Then what clearly appears to be a shock sharply breaks out the back to about 8 eV. The Dante gives a comparable signal of about 11 eV. The mystery seems to be the late time signals. The stroptometer records a "burn-thru" signal of about 20 eV at about 1.5 ns after the laser peak, whereas the Dante records a 44 eV signal emerging past 2 ns. (The Dante signal is really a flux of x-rays, whose magnitude is interpreted as a temperature by assuming that the signal emerged from the 200 μm diameter hole in the back plate of the can through which the cold side of the glass slab is visible). Besides the mystery of the 2 temperatures and 2 break out times, is the overriding mystery that the burn thru signal, if any, was expected to occur much later than either of those 2 times.

Since this was the first time such a physics experiment was attempted with laser driven hohlraums, it was important to get a fundamental understanding of the situation via simple models and rough estimates, besides doing complex 1-D and 2-D simulations with a tool like LASNEX [3]. So let us proceed systematically as we describe the data analysis.

2.2 Data analysis: Hot electron driven preheat

As mentioned above, 60 J of hot electrons with a temperature of about 70 keV were created in the primary region of the hohlraum where the laser interacts with plasma. The question is how much of this impinged onto the glass sample (we'll call it E_{inc} for energy incident). To estimate the transport of the hot electrons in the complex half-hohlraum geometry we needed to use both electron number and energy albedoes from both gold and glass of (0.5,0.7) and (0.13,0.45) respectively. Thus, for example, if 100 mono-energetic electrons impinge on glass, 13 are reflected (the rest stick) and have 45% of their initial energy. We start our crude "transport" calculation with an arbitrary 100 J worth of hot electrons in the primary. We compute the areas of the primary gold walls, the LEH, and the A_{PS}, the annular area around the cone through which the hot electrons can enter the secondary region of the half-Cairn. The electrons are then parceled out to these 3 destinations by the fraction of their areas of that total of the 3, and then the albedoes are applied. Thus after the first iteration, 38.5 % of the original 100 J worth of hot electrons are reflected from the gold wall back into the primary with 0.7 of their temperature. Another 38.5 J worth are lost in the gold wall of the primary, as is another 38.5 J x 0.3 worth (deposited energy by the reflected electrons). Another 3 J worth are lost out the LEH, and 20 J worth enter the secondary. The second iteration goes through a similar process, this time both in the primary, and in the secondary. After 3 iterations we find E_{inc} equal to 1.2 J worth of hot electrons impinging on the glass slab at their full initial temperature (70 keV), and 0.8 J at 50 keV. Had we simply taken the area ratio of the glass sample divided by the total area for the entire half-Cairn, our estimate for E_{inc} would have been 3 times higher, and wrong.

This calculation started with an arbitrary 100 J. Since our experiment had 60 J observed, we reduce our result for E_{inc} by a factor of 0.6. Moreover, since the preheat signal is overtaken by the shock signal at the time of the laser peak, only half of the (time integrated result of) 60 J contributed to the preheat signal at that time, so we multiply E_{inc} by another factor of 0.5.

We can now calculate the expected signal. Taking an ideal equation of state, but adding to the specific heat a factor of 2 to account for ionization energy, and another factor of 2 to account for the hydro expansion of the preheated material, our equation for T_e reads

$$T_e (Z + 1) = 3.8 \ 10^{-5} \ \varepsilon \ (J/gm) \qquad \text{with } Z = 0.8 \ T_{eV}^{1/2} . \qquad (1)$$

Here ε is the deposited energy due to the hot electrons in J/gm. The range λ for the hot electrons is given [4] by:

$$\lambda = 0.3 \ \mu m \ (\ T_H \ (keV) \ / \ Z_N)^2 \ Z_N^{1/2} \qquad (2)$$

where for our case of glass $Z_N = 10$, and for $T_H = 70$, 50 keV, we get $\lambda = 50$, 25 μm respectively. These ranges are considerably longer than the 12.4 μm thickness of the glass. Therefore, the ε of Eq. (1) is simply E_{inc} divided by the respective λs, and divided by the density ρ of the glass (2.5 g/cm^3) and by the area of the glass sample (7 10^{-4} cm^2). Then Eq. (1) gives T_e of 1.7 eV, in close agreement with the observed 1.4 eV.

2.3 Data analysis: X-ray driven shock waves

As mentioned in Sec. 2.1, the measure drive was $T_P = 140$ eV in the primary region of the hohlraum and, more relevantly, $T_S = 120$ eV in the secondary region which drives the glass slab. Because of length limitations on this manuscript we will not reproduce here a detailed worked example, for precisely our situation, that has already been published in reference [5]. There we worked out a model based on simple energy balance in the primary and secondary region (x-ray sources balanced by energy losses of a diffusive Marshak wave into the walls, as well as loss out the LEH and into the glass sample). Our quantitative estimates for T_p and T_S were found there to be in close agreement with the observations.

Working through the radiation ablation hydrodynamics in glass can lead to a prediction for the pressure generated by such a process: $P = 8 \ T_{heV}^3 \ t_{ns}^{-0.4}$ MB. Thus for our parameters $T = T_S = 1.2$ heV, and $t_{ns} = 1$, we expect a 14 MB shock to propagate through the glass slab. This strong shock should quadruple the pre-shock density, and thus we estimate the post shock temperature via

$$T_e (Z + 1) = (A \ / 4 \ \rho_0) \ P_{MB} \qquad \text{again with } Z = 0.8 \ T_{eV}^{1/2} \qquad (3)$$

where A is the atomic weight of glass of 20. Thus with a 14 MB shock, we expect a 9.5 eV shock temperature, which agrees well with the measured 8-11 eV. Estimating the shock velocity by the strong shock relations $v_S^2 = 4 \ P \ / \ 3 \ \rho_0$, gives $v_S = 3 \ 10^6$ cm/s. Thus the shock will traverse the 12.4 μm glass in about 400 psec, breaking out the back well within the 1 ns laser pulse width, as is indeed observed.

One dimensional LASNEX simulations of this experiment, corrected for the hot electron transport factors discussed above, indeed confirm all of our simple estimates, and reproduces the early time cold side of the glass preheat and shock temperatures. However those results also show that the cold side is not predicted to "burn-thru" and should stay at the 10 eV level. What then, accounts for the mysterious late time signals of Fig. (2)? Stepping back, and taking a larger, fresh view of the situation brought us answers.

2.4 Data analysis: X-ray driven sample motion: "Cork popping"

When the shock wave breaks out of the back of the glass, a rarefaction wave propagates back to the radiation ablation front. A pressure gradient is then set up which accelerates the "payload" – the unablated portion of the glass slab- as a whole. The central 200 μm portion of the glass is then "cookie cut" out of the hole in the back of the hohlraum, and moves out of the hohlraum. This "popped cork" then releases hot ablated gas (Fig. (3)) that was previously "bottled up" within the hohlraum and makes it visible to the 2 detectors. Thus, the glass slab indeed does not burn through, but rather allows the "cold side" detectors to see hot internal plasma as the popped cork's cold side moves out of the way of their line of sight. Since the detectors view the cold side at 2 different angles, they see a hot late-time signal at different times, corresponding to the cold popped cork moving out of their respective lines of sight.

Let us estimate the effect. The amount of glass ablated, again calculated by working through the radiation ablation hydrodynamics in glass is given by $\delta x = m / \rho_0 = 9 \ 10^{-4} \ T_{heV}^{2.25} t_{ns}^{0.6} / \rho_0$ (cm). Thus for our parameters (T= 1.2, t=1) we get $\delta x = 5.4$ μm ablated. The payload acceleration will then be given by $a = F / m = P / \rho_0 (12.4 \text{ μm} - \delta x)$ where for our case of 14 MB pressure, $a = 8 \ 10^{15}$ cm²/s. We now can calculate the distance the payload will move as:

$$d = 1/2 \ a \ t_a^2 + a \ t_a \ t = 10 \text{ μm} + 40 \ (\text{μm/ns}) \ t_{ns} \qquad (4)$$

where we take the acceleration time t_a to be 1/2 ns because the shock does not break out the back until the peak of the laser when half the 1 ns pulse is over. To estimate the late time signals of the 2 detectors we note that the stroptometer is at 55⁰ from the target normal, so the 100 μm radius, cold payload slug must travel 70 μm before the stroptometer can see past it into the hot hohlraum. The Dante is at 45⁰ so the cork must pop out 100 μm from the can before the Dante can see past its cold side into the hot hohlraum. Setting d=70, 100 μm in Eq. (4) and solving for t leads to predictions of late signals of 1.5, 2.25 ns for the 2 detectors respectively. These times are quite close to those observed as per Fig. (2). Our 1-D LASNEX simulations confirm our estimates of this payload motion as well.

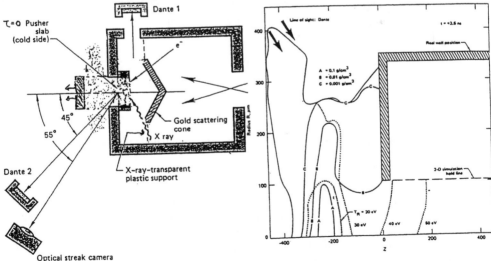

Fig. (3): schematic of cork popping.   Fig. (4): Snapshot of popped cork's ρ and T.

We are left with one remaining mystery. Why do the 2 detectors differ in the temperature they see at late times? Here 2-D LASNEX simulations help us with the answer. In Fig (4) we show a snapshot in time of the densities and temperatures of the popped cork and its surrounding hot plasma liberated from the hohlraum. The Dante line of sight passes through material of density $\rho = 5\ 10^{-3}$ g/cm^3 and temperature of 25-30 eV. The 100 eV photons detected by Dante have an opacity κ of about $3\ 10^4$ cm^2/g in such a plasma which means their mean free path, $1\ /\ \kappa\rho$, is 66 μm. Thus this hot gas liberated by cork popping is optically thick. Dante does not see all the way into the small hot exit hole, but rather (recall it is not spatially resolved) sees a much larger area radiating at 25-30 eV. Indeed as Fig (5) shows, a complete post processing of the 2-D LASNEX simulation, mimicking the Dante detector and all of its channels, shows that the emission of 100 eV photons, (dark regions in the figure) come from a large area "halo" around the cold (low emission, bright region in the figure) image of the 200 μm diameter glass plug's cold side. Moreover, in Fig. (6) we show the simulated Dante spectrum "collected" from that snapshot in time of the hot glass plasma of Figs. (4) and (5), which very closely reproduces the 5 broad Dante channels of data. Note that the spectrum has a color temperature of about 25 eV. We can integrate the energy of that spectrum, and get a total flux. If we assume misleadingly (as Dante did originally when quoting its temperature of 44 eV) that the flux is all coming from the small 200 μm diameter hole in the back plate, then LASNEX too would characterize that flux with a "brightness" temperature of 44 eV. The stroptometer is spatially resolved, so it was not "fooled", and reported the temperature of the optically thick liberated gas correctly, as 20 eV.

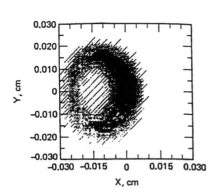

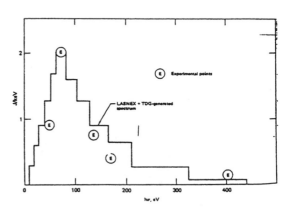

Fig. (5): Simulated cold-side image.   Fig. (6): Simulated cold-side spectrum vs. data.

2.5 Implications of the work

The shift from bewilderment to total understanding of the cold–side burn-thru signals in this, the first half-hohlraum physics experiment, gave a big boost of confidence to workers in the field, and spurred them on to make further rapid progress. Preheat detection was soon augmented by spatially resolving Kα emission diagnostics. These later helped identify B fields in hohlraum plasmas and have played a large role in fast ignitor research. Shock wave detection progressed from single slab to the more accurate stepped slab to measure shock velocities. That evolved further to wedge shaped slabs [6] to monitor shock velocity continuously, thus allowing for accurate pulse shaping data. Shocks were also produced by one x-ray driven plate colliding into another [7], acting as a power amplifier by collecting energy over a long time, storing it as kinetic energy, and delivering it rapidly back to thermal pressure during the collision. This led to very high shock pressures being achieved [8]. Further experimental refinements led to the ability to do detailed equation of state measurements [9], and material strength measurements at high pressure [10]. The cork popping was a "poor man's" way of measuring pusher speed. Soon, x-ray backlighting took over as a much clearer way to do so [11]. That soon led to quantitative measurements of hydro instabilities [12], and eventually to the mocking up of such exotic phenomenon as astrophysical jets from galactic centers[13]! While in these initial half-Cairn experiments we never really did "burn-thru", we learned enough to redesign experiments that really did have bona-fide radiation burn-thru signals, which taught us much about material opacities [5]. We used the same technique to improve hohlraum energetics by testing "cocktail" materials [14] that were combinations of elements that scattered x-rays more effectively back into the hohlraum, and burned through later in time. Eventually, exquisitely detailed frequency dependant opacities were also measured in half hohlraums as well [15]. In addition, the "ultimate" in burn-thru experiments (due to the minor role hydrodynamic motion plays in it) supersonic, diffusive radiative heat flow through low density high Z foams, driven by half hohlraums have also been achieved[16].

Thus this body of work has contributed to understanding ICF target physics and performance, and to HED science, with astrophysics as a particular application. Moreover, this body of work, already maturing by the early 90's, allowed the US national weapons lab directors to recommend that the US government attempt the experiment known as science based stockpile stewardship[17], whose most salient components are the cessation of nuclear testing, the creation of even larger HED facilities such as the National Ignition Facility, and the establishment of modern supercomputing capabilities.

3. Laboratory X-ray lasers

3.1 Solving the initial problems

By the mid 80's there had been many years of attempts at producing a soft x-ray laser in the laboratory, without success. Our first challenge was to guess what were the true reasons for those failures. The atomic physics predictions for population inversions were based on complex calculations. Could they be wrong? We took a different tack. We assumed the atomic physics predictions were reasonably accurate, and concentrated on the propagation of the x-ray laser pulse. Refraction in steep density gradients could rapidly steer the beam out of the gain medium, thus lowering the effective gain to the realm of the undetectable. We adapted a target from our ICF experience- exploding foils [2] that produced relatively uniform scale lengths, of order 100 μm, which would be sufficient to greatly lessen beam steering, and allow the effective gain to be close to its theoretical values. This turned out to be a correct guess, and a successful strategy. Analogous to the ICF work described in Sec. 2 above, we supplemented our successful full blown XRL experiments with physics experiments of the exploding foils themselves, diagnosing them in space and time, and successfully predicting that behavior, both with simple analytic models [18], and with complex LASNEX simulations.

After our initial successes on the Novette green light laser (2 beams of the Nova laser) in 1984, producing about 100 W of 200 Å radiation from 3p-3s Ne-like Se [19], we made rapid progress. By the end of the 80's we had upped that output to 1 MW [20], and demonstrated lasing at sub 100 Å with Ne – like Ag [21]. We used an analog to that scheme, the 4d-4p Ni-like scheme [22] in Ta to demonstrate gain [23] at the 45Å quite relevant for holography of biological samples in water. Our technologists developed x-ray optics, so a triple pass cavity was demonstrated [24], and initial x-ray imaging applications were accomplished as well [25, 26]. However we knew that the field would not really flourish into one filled with practical applications if the driver remained the huge 2 beams of Nova [27]. Thus, once again, we needed to step back, take a larger look at what we were doing, and find a fresh way to accomplish our goals.

3.2 Solving the next generation of problems: towards tabletop lasing

The key to our initial success was the uniform lasing medium, so we asked ourselves if we could pre-form that medium at minimal energy cost. Once formed, could we "flash heat" it to lasing conditions, using the short pulse (low energy but high power) laser technology that was becoming small, cheap, and commonplace in the late 80's. By way of comparison, the conventional way to achieve 40Å saturated (gain-length product of about 15) lasing with Ni-like W, would require a 1.5 cm long foil, with a gain coefficient of 10 cm^{-1}. That foil, in our conventional way, would be illuminated for 1 ns, at 6 10^{14} W/cm^2 with a line focussed 100 μm by 1.5 cm spot. This translates to 9 kJ – a hugely expensive proposition. If instead we simply formed the plasma with a 1 ns, 1.5 cm by 50 μm line focus spot, at the very low irradiance of 2 10^{12} W/cm^2, it could produce, over time the correct density profile that would allow propagation without refraction. That translated into a mere 15 J of energy. Then, at that proper time when the density gradient was "right", we'd bring in a short 20 ps pulse, tightly focussed to 1.5 cm by 30 μm, at the proper high irradiance of 6 10^{14} W/cm^2 to create the Ni-like state and the population inversion. That would require 54 J. All told, more than a 100 fold savings in required driver energy, and a true tabletop system. Our detailed calculations [28] of such a scheme are shown in Fig (7), and confirmed our notions of its feasibility. Note that the gain coefficient in the center of the foil is predicted to be nearly 10 cm^{-1}, as per our assumptions above. Of course it takes the XRL beam 50 ps to propagate down the full 1.5 cm length of the plasma medium, whereas the gain only lasts about 20 ps. Thus the 20 ps pump pulse will have to be "phase driven" down the 1.5 cm length of the plasma medium. There are well known techniques for doing this [29]. We also realized [28, 30] that capillary discharges could also be a good tabletop approach. The mid to late 90's featured several groups [31,32] succeeding in making tabletop XRLs using these schemes, thus making it far more likely that practical applications will one day be commonplace.

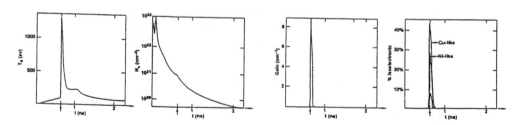

Fig. (7): At foil center, T$_e$, n$_e$, gain, & % ion-state vs. time. Arrow indicates onset of short pulse

4.0 Summary and Conclusions

The ICF/HED half-hohlraum work described in Sec. 2, and the XRL work of Sec. 3 share several features in common. Initial failures of the integrated experiment, be it high-density implosion, or actual XRL, led to designs and tests of physics experiments. Both simple models and complex simulations were important tools in those designs and in the data analysis. Solving the initial problems with those physics experiment required looking at the problem from a larger, fresh point of view. The ensuing very successful development of these fields has been most gratifying, and is a tribute to our many coworkers in these fields, which we now proceed to acknowledge.

5.0 Acknowledgements

For the half Cairn work we are grateful to: W. Mead for the initial designs, G. Tirsell and D. Banner for campaign leadership, D. Phillion, B. Price, B. Kauffman, H. Kornblum, L. Koppel, and C. Wang for diagnostic development and fielding, M. Boyle, L. Coleman and E. Storm for experimental guidance, J. Lindl and J. Nuckolls for theoretical support, B. Thomas, S. Davidson, A. Szoke, C. Alonso, and R. Ward for design/spin-off ideas, and D. Munro, B. Hammel, B. Remington, R. Collins, G. Dimonte, T. Peyser, R. Klein, J. Hammer, P. Rambo, J. Bauer, T. Back, B. Lasinski, N. Landen and P. Springer for leadership in subsequent developments.

For the XRL work we are grateful to: D. Matthews for experimental leadership, P. Hagelstein, for atomic theory / lasing scheme leadership, M. Campbell, R. London, S. Maxon and D. Eder for design support, B. MacGowan, L. Da Silva, M. Eckart, J. Trebes, N. Ceglio, R. Johnson, and G. Charatis for experimental support, A. Hazi, B. Whitten, J. Scofield, D. Lee, B. Goldstein and T. Weaver for atomic physics support, B. Hatcher, G. Rambach and G. Stone for target fabrication, and J. Rocca, J. Nilsen, S. Libby, A. Wan and J. Dunn for subsequent developments.

For computational help, and in particular, for the LASNEX simulation code support, we are indebted to: G. Zimmerman, D. Bailey, J. Harte, D. Kershaw, H. Shay, J. Larsen, M. Prasad, A. Shestakov, and E. Alley.

For general guidance and support we thank J. Nuckolls, J. Lindl, B. Tarter, W. Lokke, J. Emmett, G. Miller, D. Fortner, M. Campbell, J. Kilkenny, R. Woodruff, J. Wadsworth, M. Anastasio, C. Verdon, R. Ward, B. Goodwin, M. May, R. Batzel and E. Teller.

This work was performed at LLNL under the auspices of the U.S. DoE under contract No. W-7405-ENG-48.

References

[1] J. D. Lindl, Phys. Plasmas **2**, 3933 (1995).
[2] M. D. Rosen, et. al Phys. Rev. Lett. **54**, 106 (1985).
[3] G. B. Zimmerman and W. L. Kruer, Comm. in Plasma Physics **2**, 85 (1975).
[4] M. D. Rosen, et. al. Phys. Rev A **36**, 247 (1987).
[5] M. D. Rosen, Phys. Plasmas **7**, 1999 (2000).
[6] R. L. Kauffman, et. al Phys. Rev. Lett. **73**, 2320 (1994).
[7] M. D. Rosen, et al., in "Shock Waves in Condensed Matter -1983" J. R. Asay, G. K. Straub, and R.A. Graham, eds., North-Holland Publishing Co. (Amsterdam), 1984, pg. 323.
[8] R. C. Cauble et al., Phys. Rev. Lett. **70**, 2102 (1993).
[9] G.W. Collins et al., Science **281**, 1178 (1998).
[10] D. H. Kalantar, et al., Phys. Plasmas **3**, 1803 (1996).

[11] R. H. Price, et al., in "Shock Waves in Condensed Matter - 1981" W. J. Nellis,L. Seaman, and   R. A. Graham, eds., American Institute of Physics **78**, 155 (1982).

[12] J. D. Kilkenny, Phys. Fluids B **2**, 1400 (1990).

[13] B. A. Remington, et al., Science **284**, 1488 (1999).

[14] T. J. Orzechowski, et al., Phys. Rev. Lett. **77**, 3545 (1996).

[15] P. T. Springer, et al., Phys. Rev. Lett. **69**, 3735 (1992).

[16] C. A. Back, et al., Phys. Rev. Lett. **84**, 274 (2000).

[17] R. Jeanloz, Physics Today **53**, 44 (2000).

[18] R. A. London and M. D. Rosen, Phys. Fluids **29**, 3813 (1986).

[19] D. L. Matthews, et al., Phys. Rev. Lett. **54**, 110 (1985).

[20] B. J. MacGowan, et al., Phys. Fluids B **4**, 2326 (1992).

[21] D. J. Fields, et al., Phys. Rev. A **46**, 1606 (1992).

[22] M. S. Maxon, et al., Phys. Rev. Lett. **63**, 236 (1989).

[23] B. J. MacGowan, et al., Phys. Rev. Lett. **65**, 420 (1990).

[24] N. M. Ceglio, et al., Opt. Lett. **13**, 108 (1988).

[25] L. B. Da Silva, et al., Science **258**, 269 (1992).

[26] D. M. Ress, et al., Science **265**, 514 (1994).

[27] D. C. Eder, Phys. Plasmas **1**, 1744 (1994).

[28] M. D. Rosen, OSA Proceedings on Short Wavelength Coherent Radiation: Generation and Applications, P. Buchsbaum and N. Ceglio eds. (OSA, Wash. DC 1991) Vol. 11, pg. 96.

[29] Z. Bor, et al., Appl. Phys. B. **32**, 101 (1983).

[30] M. D. Rosen and D. L. Matthews, U.S. Patent # 5016250 (5/14/1991).

[31] J. J. Rocca, et al., Phys. Rev. Lett. **77**, 1476 (1996).

[32] J. Dunn, Phys. et al., Phys. Rev. Lett. **84**, 4834 (2000).

LARRY SUTER
2003
(with co-authors)

Prospects for high-gain, high yield NIF targets driven by 2ω (green) light[*]

L. J. Suter, S. Glenzer, S. Haan, B. Hammel, K. Manes, N. Meezan,
J. Moody. M. Spaeth *Lawrence Livermore National Laboratory,
University of California, Livermore, CA 94551*
K. Oades, M. Stevenson *AWE*

For several years we have been exploring the possibility of using green (2w) light for indirect drive ignition on NIF. The rationale for this work is the possibility of extracting significantly more energy from NIF in green light, as compared to blue (3w) light, and driving far more energetic capsules than we originally envisioned when we started planning NIF in the early 1990's. This paper attempts to provide a comprehensive picture of the progress we have made exploring 2w for NIF ignition. First we describe the potential operating regime for NIF at 2w and how that can translate into a very large "design space" for exploring ignition target designs. We then present the results of 2w ignition target design studies indicating that we can achieving adequate drive and symmetry with 2w and showing how we might capitalize on the large amount of energy available by electing to trade-off coupling efficiency for, say, better symmetry or plasma conditions. These simulations also define plasma conditions for ignition-relevant 2w laser-plasma interaction experiments that have been recently performed. We summarize the results of these experiments which indicate that 2w LPI is not very different from 3w's. Finally, we show how recent experimental findings on mitigating 2w laser plasma interactions through reduced intensity and/or judicious choice of plasma composition can be incorporated into ignition target designs.

Section I- Potential target design space available with 2w

The fundamental requirements of the National Ignition Facility (NIF) Laser now being constructed at Lawrence Livermore National Laboratory

---

[*] Submitted for the Proceedings of the 3rd International conference "INERTIAL FUSION SCIENCE AND APPLCATIONS, IFSA2003" to be published by the American Nuclear Society (2004) B.J. Hogan et al eds.

include that it be capable of irradiating a target with 1.8MJ of 0.35μm (3w or "blue") light in an ignition pulse shape peaking at 500TW. The 3w light is produced by a neodinium phosphate glass laser system [1] which first produces infrared or "1w" light of 1.053μm wavelength which is then converted to 3w light by a pair of KDP crystals [2]. The crystals combine three 1w photons into one 3w photon. Ignition pulse shapes require peak power after a long, low power "foot" lasting many nanoseconds. Moreover, this peak power must be produced after a significant amount of energy has already been extracted from the 1μm laser since crystals tuned to provide optimum, ~70% conversion of 1w to 3w at peak power have

relatively low conversion during the foot. Since the average conversion efficiency for a 3w ignition pulse shape, without any advanced conversion schemes, works out to be about 50%, NIF's "1μm engine" was designed to produce ~700TW of 1w power after >3MJ of 1μm energy has been extracted. The consequence of this is

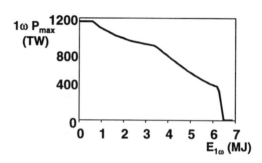

Figure 1- NIF's estimated maximum 1w output power vs. 1w extracted energy.

that NIF's 1μm laser is necessarily very much larger than the 1.8MJ specified output. Figure 1 plots current estimates of NIF's maximum potential for producing 1w energy and power. It indicates that NIF's peak, extractable 1w energy would be ~6.5MJ. This estimate is for NIF's so-called 11-7 configuration with all seven "slots" in the final, booster amplifiers loaded with slabs of neodynium glass. We note that at this writing the first four beams of NIF have already demonstrated [3] 104kJ of 1μm light output, equivalent to 5MJ of 1w from full 192 beam NIF.

Previously [4] we discussed how 3w operations at lower powers, in tandem with improvements in hohlraum coupling efficiency, might allow NIF to drive capsules ~400-600kJ. In this paper we present an assessment of the possibilities offered by operating NIF as a green, 2w laser and show how it allows ignition and, even, high yield opportunities far beyond what we originally envisioned when we started NIF in the early 90's. NIF's potential for driving ignition targets with 2w can be simply estimated by $E_{cap}=E_{1\omega}*(\eta_{1-2})*\eta_H$, where E_{cap} is the capsule absorbed energy, $E_{1\omega}$ is NIF's maximum 1w output (~6.5MJ), η_{1-2} is the conversion efficiency of 1w laser energy to 2w (~80-85% average conversions to green are possible) and η_H is the hohlraum coupling efficiency. This gives E_{cap}~5MJ*η_H, or capsules absorbing >~1MJ energy at plausible hohlraum coupling efficiencies of 20-25% [4].

Figure 2 summarizes a more detailed analysis and graphically shows NIF's potential "ignition target design space" with 2w. It plots NIF energy vs. capsule absorbed energy. The light and moderate shaded areas show target design space available with 2w at 250eV peak radiation temperature and 300eV peak radiation temperature, respectively. We refer to this as

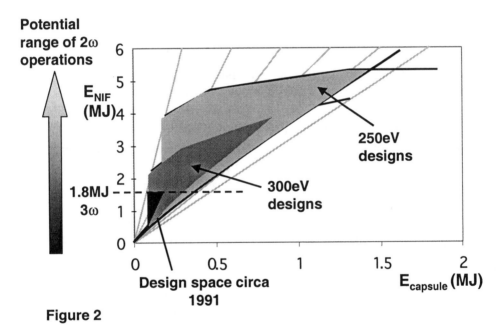

Figure 2

target design spaces because it illustrates all the combinations of NIF energy and capsule energy where an ignition target might be designed at a given peak radiation temperature. Both these design spaces are very much larger than the design space we originally envisioned in ~1991 when we began NIF, shown by the dark triangle in the lower left section.

To better appreciate the target design space plots of figure 2 and to understand how they are developed we begin by noting that the light grey, straight lines are lines of constant hohlraum coupling efficiencies, η_H. The bold lines bounding the right hand sides of the 250eV and 300eV design spaces are estimates of coupling efficiency for cylindrical NIF ignition hohlraums with a "standard" case:capsule ratio=(hohlraum area/capsule area)$^{1/2}$ of 3.65 [4]. These efficiencies have a slight non-linearity, $\sim(E_{cap})^{0.1}$. The left hand, vertical boundaries indicate estimated minimum energy of ignition at a give peak radiation temperature. The boundaries drawn combine the approximate minimum energy of ignition at 300eV, generally taken to be $\sim$100kJ, and the $T_R^{4.5}$ scaling for minimum energy developed by Lindl, assuming "similar" targets. We note, however, that the minimum energy for ignition can be significantly affected by target design and is the subject of ongoing research. For example Dittrich [5] has designed a capsule which, at 250eV, also has a minimum energy of ignition of $\sim$100kJ.

The upper bound of target design space is found at each hohlraum coupling efficiency by combining target pulse shape requirements with the conversion efficiency, η_{1-2}, of 1w to 2w light and NIF's 1w performance curve, figure 1. The procedure is as follows: Target pulse shape requirements are derived from the x-ray power vs. time absorbed by

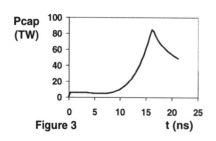

Figure 3

a given capsule. Figure 3 shows the x-ray power absorbed by a 600kJ, 250eV graded dopant Be capsule designed by Haan in 1991 (designated "Haan'91") [4]. We readily scale this capsule's x-ray power requirements, P_{cap0}, to other absorbed energies via a capsule scaling parameter "s". Multiplying P_{cap0} by s^2, time by s and all the dimensions of the capsule by s scales capsule absorbed energy by s^3 or Ecap=600kJ$*s^3$ for this scaled capsule. For a given Ecap the 2w pulse shape requirement is simply (x-ray power absorbed by the capsule)/(hohlraum coupling efficiency)= P_{cap}/η_H. The 1w power produced by the laser must then be $P_{1w}=P_{cap}/(\eta_H \eta_{1-2})$. However, figure 1 provides a constraint on the maximum 1w power, requiring $P_{1\omega} = P_{cap}/(\eta_H \eta_{1-2}) < Pmax(E_{1\omega})$ where $E_{1\omega} = \int_0^t P_{1w}dt$. To find an upper bound at a given η_H and η_{1-2} we vary the capsule scale size parameter, s, until there is a point in the pulse shape where $Pcap/(\eta_H \eta_{1-2}) = Pmax(E_{1\omega})$.

Figure 4 plots these upper bounds for 250eV and 300eV peak radiation temperatures for 1ω to 2ω conversion efficiencies, η_{1-2}, ranging between 50 and 90%. Note that the upper bounds have a significant dependence on peak radiation temperature and on 1ω to 2ω conversion efficiency. The notable difference between the 300eV and 250eV upper bounds is because 300eV capsules require about twice the power of the 250eV capsules. 300eV targets are always power limited. It is not unreasonable to think of 250eV as marking the approximate boundary between designs limited by available power and ones limited by available energy. Analysis at 215eV, approximately a factor of 2 down in power requirement from 250eV, shows a design space only a bit larger than the 250eV space. At 215eV the targets are mostly limited by the 1w energy available (giving a flat upper bound in the E_{NIF} E_{CAP} plot), except at the lowest hohlraum coupling efficiencies where they, too, become power limited. We also note that at 300eV, 50% 1ω to 2ω conversion efficiency, the design space is not very much larger than the one we originally envisioned for 300eV NIF targets.

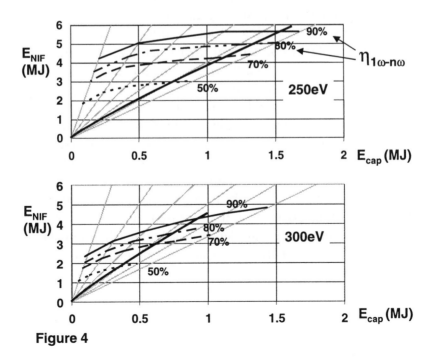

Figure 4

Figure 2 shows an ignition target design space using 2ω that is far greater than the target design space that existed in the early '90s, when we first started thinking about NIF. At 300eV the increase comes principally from increased conversion efficiency, an increase in our expected coupling efficiency [4] and a clearer understanding of how the 1w laser will operate. Further expansion of target design space comes from operating at 250eV rather than 300eV, where the power requirements as a function of extracted energy are better matched to NIF's 1w capabilities. In order to achieve this performance considerably more green energy must pass through NIF's final optics assembly (FOA) than the 8J/cm^2 of blue light that will pass through the FOAs during a 1.8MJ 3w shot. Although this fluence currently defines the state-of-the-art for 3w optics it is expected that much higher fluences will be possible with green light. Current thinking is that if full NIF were available today it could reasonably produce between 3 and 4MJ of green light. With further optics research it is conceivable that 2w optics could allow access to the entire design space.

Section II- Discussion: Benefits of larger design space and 2w target physics concerns

The preceding section showed that NIF, operating at 2w, has the potential to greatly increase target design space compared to our original expectations. This increase is desirable for several reasons. First, it allows us to contemplate capsules absorbing far more energy than we originally envisioned. Capsules absorbing ~100kJ (200kJ) are on the threshold of failure at 300eV (250eV) because of their small size [6,4]. Basically, as a given ignition capsule is scaled down in size and energy, heat conduction losses play an increasing dominant role in the hotspot power balance, causing 1-D estimates of yield vs. absorbed energy to have a very steep section or "cliff" at low energies. Significantly increasing capsule absorbed energy moves us away from this cliff. Increased capsule absorbed energy is also beneficial because a given capsule's ability to withstand surface roughness, which seeds hydrodynamic instabilities, increases very dramatically with absorbed energy [7]. Such important increases in margin, possible with increased capsule absorbed energy, would greatly increase our confidence in achieving ignition and allow us to consider studies of capsule physics and thermonuclear burn physics that are implausible with marginal capsules. A second reason the increase in design space is attractive is that it allows us to consider a wider range of possible hohlraums and to consider the possibility of trading-off capsule absorbed energy for something desirable such as better symmetry or improved diagnostic access.

The increase in target design space potentially available with 2w makes it appear to be a desirable option for NIF. Unfortunately, virtually all the target physics studies that established the technical basis for NIF ignition [6,8] were done with 3w light. When we first recognized the possibilities of green light no significant 2w database existed. In order to redress this we have been working for several years to answer questions related to using 2w light on NIF for ignition. This work has been divided into three major areas.

1- Laser operations: What performance might we get from NIF at 2w and how might we actually operate NIF at 2w. The previous section described the ignition performance we might get with 2w. Work assessing 2w operations is ongoing within the NIF project and will be reported elsewhere.

2- Projected 2w ignition target performance assuming Laser Plasma Interactions are under control. In the next section we describe the result of integrated Lasnex simulations of large 2w ignition targets.

3- Experimental studies of Laser Plasma Interaction issues for which there is no theoretical predictive capability. The key issues are 2w propagation, 2w backscatter and 2w hot electron production. We have been studying these on both the Helen laser and Omega laser. We summarize current findings in section IV.

Section III- Lasnex studies of 2w ignition targets

The large target design space potentially available with 2w light gives us the luxury of being able to consider a wide range of ignition possibilities with Lasnex. Referring to the 250eV design space of figure 2, we have done integrated Lasnex simulations [9] of two targets that require ~3.5 MJ of green light. One target, with a standard case:capsule ratio of 3.65, lies on the limiting hohlraum coupling efficiency line, driving a capsule that absorbs ~850 kJ of x-rays. A second target demonstrates one of the trade-offs made possible by a very large target design space. It contains a capsule that absorbs only 400 kJ of x-rays, allowing us to increase the case:capsule ratio to 5.

Figure 5 shows the two targets we simulated with Lasnex. The target with standard, 3.65 case:capsule ratio has a Be "Haan'91" capsule ~4mm outer diameter placed inside an ~1cm diameter cocktail hohlraum [4]. This hohlraum is ~1.74X the size of the NIF point design [10]. The capsule absorbs 850kJ. In the other target is a scaled version of the same capsule with a diameter of ~3mm. It is inside a cocktail hohlraum ~1.1cm in diameter (scale 1.93), giving a case:capsule ratio of 5.0 and a capsule absorbed energy of ~400kJ. The 2w pulse shapes we used in simulating the two targets are plotted in figure 6. Both are continuous pulse shapes of approximately 3.5 MJ. The target with larger case:capsule ratio requires higher power because its smaller capsule implodes more quickly; the ~same amount of energy must come in a shorter time. As in the original point designs, both targets include a low-Z gas fill (1mg/cc He) to retard the inward motion of the high-Z hohlraum walls in order to maintain symmetry [10,11]. We used typical NIF beam pointing as originally developed by Pollaine [10]. A variety of spot sizes were explored in the integrated simulations, including spots as large as ~4mm major diameter by ~1.5mm (3mm) minor diameter for the 44.5°&50° (23°&30°) beam cones. Spots this large are closely matched to the laser entrance hole (LEH) size thereby minimizing intensity at

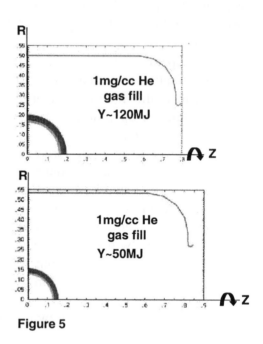

Figure 5

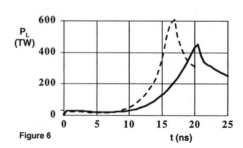

Figure 6

the LEH. These "big spot" integrated simulations give results very similar to simulations with considerably smaller spots. Using the large spot size with the 850kJ capsule's pulse shape gives a peak single-quad intensity of ~3×10^{14} w/cm^2 (~1.5×10^{14} w/cm^2) for quads on NIF's outer (inner) cones.

Extensive, 2D integrated Lasnex simulations indicate 2w is very promising for igntion. The calculations, using the large spots just described, produce the desired $T_R(t)$ in the hohlraum. Indeed when we perform an identical simulation, except replacing 2w with 3w, we find nearly identical $T_R(t)$ profiles. The small differences can be attributed to slightly higher temperature of the hohlraum's coronal plasma with 2w. We find that the simulated 2w beams propagate to the walls and that we can control symmetry in the usual way, by moving the beams and/or adjusting the relative powers [10,11]. Consequently, we produce adequate symmetry and the capsules ignite and burn in our 2w integrated simulations. The 850kJ capsule produces ~120 MJ and the 400 kJ capsule produces ~50 MJ. Both these yields are close to the 1D yields for these particular capsules driven by idealized $T_R(t)$ pulse shapes.

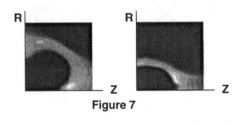

Figure 7

Figure 7 illustrates hotspot shape at ignition time for the two case:capsule ratios. We define ignition time as when the thermonuclear yield rises through ~2000TW, a useful rule-of-thumb criteria. At standard case:capsule ratio we see a hotspot shape (left image) that is very familiar from 3w design work; a well formed hotspot showing evidence of an incipient jet of cold DT on the pole together with an incipient curtain of cold DT coming in around the waist of the capsule. Neither perturbation is sufficiently large to affect ignition (indeed, we find at these large absorbed energies that many poorly tuned targets with very much bigger jets and/or curtains will also ignite on the code). At larger case:capsule ratio (right image) we see evidence of a trade-off worth further exploration. The symmetry appears to have been improved. This hotspot shows no evidence of a budding pole jet or waist curtain. In tuning the symmetry we find that with these bigger capsules we can achieve adequate symmetry with out needing time dependent "beam phasing". That is, without continuously and carefully varying the ratio of inner beam power to outer beam power to minimize time dependent variations in the P2 legendre component of asymmetry [10,11]. For a given pointing, one, fixed in time ratio is adequate. That is not to say that with increased coupling energy we still wouldn't want to try to improve symmetry via some beam phasing. The fact that we don't necessarily need to use beam phasing in successful integrated simulations is an anecdotal measure of increased robustness due to increased absorbed energy.

The weakness of the design simulations just discussed is that neither Lasnex nor any other model can quantitatively predict LPI. In creating the technical basis for NIF we dealt with this shortcoming by doing a wide variety of Nova underdense interaction studies [12, 6, 8] in targets we considered to be "ignition plasma emulators". That is, targets in which we had created, to the degree possible, plasma conditions similar to what we

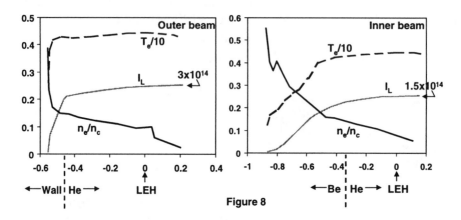

Figure 8

expect in ignition targets. Lasnex integrated simulations of ignition targets defined those plasma conditions. The plasma conditions from our integrated simulations of the 250eV ignition target at 3.65 case:capsule ratio, at 1ns after peak power are plotted in figure 8 for the inner and outer cones. According to these plots, LPI for the outer beam principally involves a beam of $\sim 3 \times 10^{14}$w/cm^2 interacting with a low-Z plasma with $T_e \sim 4$keV and electron density ~ 0.1 to $0.14 n_c$, where n_c is the critical density for green light, 4×10^{21} electrons/cm^3. For the inner beams, LPI occurs at a lower intensity, $\sim 1.5 \times 10^{14}$w/cm^2, and in a plasma that changes from He fill-gas to Be ablator blow-off about midway in the beam's path. The plasma density along this path ranges from ~ 0.1 to $0.2 n_c$. These conditions, then, determine the conditions for empirical studies of laser

plasma interactions in a 2w ignition target and how we might control them.
Section IV- Experimental studies of 2w Laser Plasma Interaction (LPI)

The key underdense interaction issues for 2w are essentially the same as they are for 3w; propagation, backscatter and hot electron production. For 3w ignition these issues were studied on the Nova laser during the 1990's as part of the Nova Technical Contract [6,8] that created the target physics basis for ignition with a NIF class facility. Of these issues, backscatter losses were the greatest concern and were studied in depth while hot electron production, which had never been observed to be large with 3w, was monitored on Nova but never became the focus of detailed experiments.

In order to establish a database for laser plasma interactions at 2w we have been studying underdense interactions on the 2w Helen laser at AWE [13] since 2000 and have converted one beam of the Omega laser at University of Rochester to operate at 2w. We have been shooting green interaction experiments at Rochester since June, 2002.

The 2w Omega experiments have principally studied backscatter and are described in greater detail by Moody [14]. Conceptually, the Omega studies are very similar to underdense interaction experiments carried out on Nova to study 3w. They use a so-called "gasbag" target comprised of two thin (~3500A) polyimide membranes glued to either side of an aluminum washer which also has tiny tubes for filling the target with gas. When pressurized, the membranes stretch, forming an oblate spheroid of major diameter ~2.75mm and minor diameter ~2.2mm. These gasbags are heated by 1ns pulses from 40 of Omega's beams. The heater beams are defocussed to low intensity, nearly filling the bag's diameter and create a plasma with Te~ 2.5keV and scalelength >1mm. This plasma then "probed" by Omega's single 2w beam which has ~400J in a 1ns pulse. The 2w probe beam is smoothed by a phase plate which forms a spot that can achieve intensities up to ~1x10^{15}w/cm^2. Backscatter into the f/6 lens, the principal quantity studied, is measured by Omega's Full Aperture Bacscatter Station (FABS). At this point, Omega does not yet have an "NBI" diagnostic to measure 2w light scattered just outside the lens. Figure 9 shows one of the scalings we have performed on Omega. It plots 2w Raman and Brillouin reflectivity as a function of intensity from gasbags filled with hydrocarbon gas to a density corresponding to 0.12nc of green light when the gas is fully ionized. We make several observations from this plot. First, hydrocarbon gasses at 2w, like 3w, mainly produce Raman backscatter at 0.12 nc. Second, the peak Raman backscatter into the lens at intensity approaching 10^{15} is ~15%, a typical value for 3w light at similar conditions. Third, there is a clear intensity scaling to the backscatter that could be interpreted as a threshold for Raman at low-10^{14} w/cm^2.

A threshold for Raman backscatter at low-10^{14} w/cm^2 is encouraging because it can be explained by a filamentation argument and support for that explanation can be found in the data. At the heart of the filamentation argument is an assumption that Raman backscatter is produced mostly in the hotspots that form when the beam filaments. That is, if the beam filaments we get Raman backscatter but if the beam doesn't filament then Raman should be low. In these experiments simple theory and simulations with our laser plasma interaction code pF3D [15] indicate a threshold for filamentation around 3×10^{14} w/cm^2. This threshold is supported experimentally by a very narrow Raman backscatter spectrum at 3×10^{14} but very obviously broad Raman backscatter spectra at the higher intensities. (Broad Raman spectra are indicative of filamentation while a narrow spectrum is indicative of little or no filamentation). If the filamentation threshold hypothesis is correct, then this scales favorably to 2w NIF ignition targets since the intensity threshold for filamentation should scale like ~Te [16]. In Lasnex simulations of 2w ignition targets Te is ~5keV, vs. ~2.5keV in the Omega experiments.

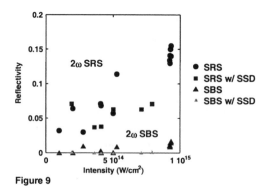

Figure 9

Complementing 2w interaction experiments on Omega have been a wide ranging series of underdense interaction experiments using a single, ~400J/1ns 2w beam on the Helen laser at AWE. The experiments mostly involved gasbag targets irradiated along the axis of symmetry by a phase-plate smoothed, 2w spot, typically ~6x10^{14} w/cm^2. A number of small, gas filled hohlraums were also shot, as well. The experiments are described in detail in a paper by Stevenson [17]. Here we summarize the three most important Helen findings on underdense interaction.

1- Propagation: Because the gasbag targets were irradiated by a single beam, we were able to study 2w propagation via time resolved, side-on x-ray imaging. These side-on images show the formation of well defined plasma columns and closely match synthetic images from simulations with our radiation hydrodynamics code HYDRA [18]. We interpret the good agreement between experimental and synthetic images as evidence that a 2w beam can propagate in a manner consistent with straightforward hydrodynamics and evidence that, for backscatter production, these targets produced the long scalelength plasmas we expected from simulations [19].

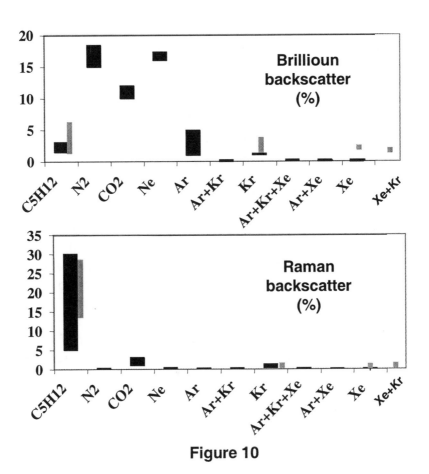

Figure 10

2- Effect of composition on backscatter: Helen's seminal contribution to LPI is the discovery that plasma composition influences backscatter far more than was previously thought [17]. The wide bars on figure 10 show Raman and Brillouin backscatter from the Helen gasbag targets as a function of composition, ordered in increasing average atomic number, Z. The vertical extent of a bar indicates the range of backscatter we measured from all targets filled with a given composition. At low Z we see the expected interchange of Raman for Brillouin when we switch from a composition with strong ion damping (C_5H_{12}) to a composition with weak ion damping (N_2, CO_2, Ne). This is consistent with Nova result [20]. The unexpected finding was the drop in Brillioun with rising Z and the fact that Raman remained low even as Brillouin dropped. This was inconsistent with a widely held view of an interplay between Raman and Brillouin and that reduction of one results in the increase of the other. These findings have been reproduced in subsequent Omega gasbag experiments using 40 heater beams and one probe. The narrow bars plot the Omega results.

3- Control of hot electron production: Historically, hot electron production was the bane of early attempts to do ICF with lasers having wavelengths of 1μm or longer. For example, experiments on the 1μm Shiva laser showed hot electron production to rise as hohlraums are driven to higher energy density and that in the highest energy density hohlraums it was possible to convert >20% of the laser energy to hot electrons with an ~50keV maxwellian distribution. Such high hot electron fractions prevent ignition by preheating the DT fuel. The discovery in the early '80's that shorter wavelengths suppress hot electron production led the community to build subsequent facilities to operate at the shortest wavelength technically feasible, 3w. Long experience has justified that decision. Empirically, 3w does not efficiently make hot electrons. When considering the possibility of using 2w, history cautions to be wary of the specter of hot electrons. This where Helen experiments have made a second original contribution to LPI; hot electron production and how to control it. Measurements of time integrated, absolute hard x-ray production with Helen's Filter Fluourescer diagnostic (FFLEX) allow us to infer hot electron production. In gasbag targets, we find that C_5H_{12} fills, which efficiently produce Raman backscatter, also produce a rising hot electron fraction as a function of fill density. However, when we switch to other gasses, which do not produce much Raman, the hot electron signal remains relatively low, even when as the fill density approaches 0.25nc. This indicates that plasma composition can control hot electron production, just as it appears to control backscatter. Complementing the gasbag experiments, we also shot some small, gas filled gold hohlraums on Helen in order to further study hot electron production. These experiments used unsmoothed beams, with best focus (~80µm diameter) at the LEH. Figure 11 plots hot electron fraction observed with these hohlraums as a function of fill density, for two fills, C_5H_{12} and Kr. With C_5H_{12} there is a striking increase in hot electron production with fill density with a peak, inferred hot electron fraction of ~20% in the vicinity of $0.25n_c$. However, when we change the fill gas to Kr, we find relatively little hot electron production, even near $0.25n_c$. Backscatter measurements on these hohlraums also show that Raman is relatively high in the C_5H_{12} hohlraums but very much reduced in the Kr filled hohlraums. These Helen experiments suggest two rules-of-thumb for designing 2w ignition targets with low hot electron production. Keep most of the LPI volume below 0.15nc and/or judiciously choose materials to avoid Raman producers.

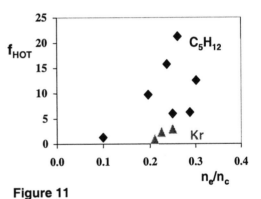

Figure 11

Section V- Alternative ignition hohlraum designs

The finding that we can control backscatter and hot electron production via judicious choice of plasma composition is potentially very important for NIF because it implies that we can control LPI via target design. This has engendered a new area of target design, exploring targets where the conventional He/H gas-fill of an ignition point [10,11] is replaced by other materials. A constraint on these designs is a preshot temperature of ~18°K needed for the cryogenic capsule. This eliminates most gasses since they would freeze out.

Our exploration of alternative hohlraum designs has been exclusively on variants of the standard case:capsule ratio target of figure 5, using the pulse shape shown by the solid line of figure 6 and the 850 kJ Be capsule. Our investigations fall into two cryogenic-compatible classes; designs where the He gas fill is replaced by a foam and designs where it is largely replaced by a mid-Z or high-Z liner. In the foam designs we replaced the 1mg/cc He gas by 1mg/cc SiO_2 (this foam exists) or 1mg/cc GeO_2 and, even, 1mg/cc XeO_2 (this foam cannot exist but allows us to study the scaling with Z). The lined targets we studied included hohlraums lined with 1μm solid (frozen) Kr and 1μm frozen Xe.

The result of these integrated simulations is that it appears possible to replace the He or He-H gas in NIF hohlraums with mid to high Z material and still maintain drive and symmetry adequate for ignition. The calculated $T_R(t)$ from hohlraum simulations using the three different foams are close to what is calculated for He fill. The simulated hotspot shapes at ignition do not look very much different that the one found with He fill, in figure 7. The capsules work in these integrated simulations, producing yields ~120MJ, similar to He filled targets. It was not necessary to make any design changes compensating for the increased x-ray preheat of the higher-Z foams. The principal drawback is that the hohlraum has a greater propensity to produce a pole high implosion as we raise the average Z of the fill. In this study we counteracted this tendency by switching a greater fraction of the laser power to the inner beams. If the pole-high tendency cannot be offset by some other change, such as geometry, this might limit the upper bound to the Z of the foam.

In addition to the foam simulations, we also investigated replacing the He gas with 1μm liners of either Kr or Xe. Although these designs readily produced the required $T_R(t)$, we were unable to find a symmetry solution for vacuum hohlraums with liners. Axial stagnation of the liner material at later times generated a pole-high x-ray pulse that could not be offset by raising the power of the inner beams. However, if we included a very low fill of He, 0.1mg/cc, we found we could tune the symmetry. In these simulations we again found it necessary to raise the fraction of power to the inner beam in order to tune the symmetry.

These few preliminary simulations of alternative hohlraums are far from being detailed point designs. However, they do show that it is possible to consider replacing the He or He-H gas of the conventional NIF designs with some other material and still be able to produce the required drive and adequate symmetry for ignition. This, in turns, means that it could be possible to engineer LPI in 2w (or, even, 3w) ignition designs by engineering the plasma composition in the beam paths. Alternative hohlraums are a new area of investigation that we will be examining in the coming years.

This work performed under the auspices of the U. S. Department of Energy by the Lawrence Livermore National Laboratory under Contract No. W-7405-Eng-48.

References:

1- J. A. Paisner, E. M. Campbell and W. J. Hogan, Fusion Technol. **26**, 755 (1994)

2- R. S. Craxton, "Theory of high-efficiency third harmonic generation of high-power Nd-glass radiation." Opt. Commun. 34, 474-478 (1980).; B. M. Van Wonterghem et. al. , Appl. Opt. 36, 4932 (1997).

3- K. Manes, M. Spaeth, LLNL, private communication, 2003.

4- L. Suter, J. Rothenberg, D. Munro, B. Van Wonterghem and S. Haan, Phys. Plasmas 7, 2092 (2000).

5- T. Dittrich, S. W. Haan, M. M. Marinak, S. M. Pollaine, and R. McEachern, Phys. Plasmas 5, 3708 (1998).

6- J. D. Lindl, Phys. Plasmas **2**, 3933 (1995).

7- S.W. Haan, T. Dittrich, G. Strobel, S. Hatchett, D. Hinkel, M. Marinak, D. Munro, O. Jones, S. Pollaine, and L. Suter, "Update on ignition target fabrication specifications," *Fusion Science and Tech.* **41**, 165 (2002); G. Strobel, S. W. Haan, et. al. Submitted for publication.

8- "Special Issue: *Nova Technical Contract"*, ICF Quarterly Report, July-September 1995, p. 209 (UCRL-LR-105821-95-4) (unpublished).

9- L. J. Suter, A. A. Hauer, L. V. Powers, et. al., Phys. Rev. Lett. 73, 2328 (1994).

10- S. W. Haan, S. M. Pollaine, J. D. Lindl et. al., Phys. Plasmas 2, 2480 (1995).

11- W. J. Krauser, N. M. Hoffman, D. C. Wilson, et. al., Phys. Plasmas 3, 2084 (1996).

12- B. J. MacGowan, B. B. Afeyan, C. A. Back, R. L. Berger, Phys. Plasmas 3, 2029 (1996).

13- M.J. Norman, J.E. Andrew, T.H. Bett *et al,* Appl. Opt. **41** , 3497 (2002)

14- J. Moody, et. al., Proceedings of the Third International Conference of Inertial Fusion Sciences and Applications, Monterey, CA, 2003 (to be published).

15- R. L. Berger, B. F. Lasinski, T. B. Kaiser, E. A. Williams, A. B. Langdon, and B. I. Cohen, Phys. Fluids B **5**, 2243 (1993); C. H. Still, R. L. Berger, A. B. Langdon, and E. A. Williams, "Three-dimensional nonlinear hydrodynamics code to study laser-plasma interactions," ICF Quarterly Report, July–September 1996, p. 138 (UCRL-LR-105821-9-4) (unpublished).

16- E. A. Williams, LLNL, private communication (1999).

17- M. Stevenson, Phys. Plasmas, 2004 (to be published)

18- M. M. Marinak, G. D. Kerbel, N. A. Gentile, *et al,* Phys. Plasmas, **8**, 2275 (2001)

19- N. Meezan, et. al., Proceedings of the Third International Conference of Inertial Fusion Sciences and Applications, Monterey, CA, 2003 (to be published).

20- D.S. Montgomery, B.B. Afeyan, J.A Cobble *et al*, Phys. Of Plasmas,.**5**, 1973 (1998)

HIDEAKI TAKABE
2003

Hydrodynamic Instability, Integrated Code, Laboratory Astrophysics, and Astrophysics

Hideaki Takabe

Institute of Laser Energetics, Osaka University
Suita, Osaka 565-0871 JAPAN
takabe@ile.osaka-u.ac.jp

Abstract: This is an article for the memorial lecture of Edward Teller Medal and is presented as memorial lecture at the IFSA03 conference held on September 12th, 2003, at Monterey, CA. The author focuses on his main contributions to fusion science and its extension to astrophysics in the field of theory and computation by picking up five topics. The first one is the anomalous resistivity to hot electrons penetrating over-dense region through the ion wave turbulence driven by the return current compensating the current flow by the hot electrons. It is concluded that almost the same value of potential as the average kinetic energy of the hot electrons is realized to prevent the penetration of the hot electrons. The second is the ablative stabilization of Rayleigh-Taylor instability at ablation front and its dispersion relation so-called Takabe formula. This formula gave a principal guideline for stable target design. The author has developed an integrated code ILESTA (1D & 2D) for analyses and design of laser produced plasma including implosion dynamics. It is also applied to design high gain targets. The third is the development of the integrated code ILESTA. The forth is on Laboratory Astrophysics with intense lasers. This consists of two parts; one is review on its historical background and the other is on how we relate laser plasma to wide-ranging astrophysics and the purposes for promoting such research. In relation to one purpose, I gave a comment on anomalous transport of relativistic electrons in Fast Ignition laser fusion scheme. Finally, I briefly summarize recent activity in relation to application of the author's experience to the development of an integrated code for studying extreme phenomena in astrophysics.

1. Introduction

It is great honor for me to be awarded this distinguished Medal. As you know, Edward Teller is mostly known as the very first to produce an exothermal nuclear fusion reaction. I may reflect this after I was invited to the 3rd Sakharov Physics Conference in Moscow in June 2002. Sakharov is the counter-person in the former Soviet Union. What I was surprised after listening to all the plenary sessions of Sakharov physics conference is that his scientific field is extremely wide. He is famous as a pioneer of cosmology, and the spectral oscillation of the 2.7K cosmic microwave background fluctuations is called "Sakharov oscillation". I thought he is an outstanding genius of physics.

This is also true for the case of US counter-person, Edward Teller. I know his footprint in atomic physics with Inglis-Teller limit, in nuclear physics with Gamov-Teller transition, in Quark physics with Quark Gluon Plasma, and so on. He was at least a similar outstanding genius of physics as Sakharov.

Almost forty years have passed since the invention of laser. It is usually said that the laser is one of the greatest inventions in the 20th Century. In some case, an invention leads to an unexpected development with pressure from society's needs or with progress of science. I find that the imagination of human beings is wonderful by the fact that people had the dream to achieve the laser fusion by increasing the output of the laser by 100 thousand times. This was just at the time (1962) when a pulsed laser could be made in the laboratory. This story encourages us when we have some unusual idea. I am, however, not so naive as to believe that only such a dream of a scientist has led to the construction of the present-day huge laser systems. That achievement occurred because of the strong or even hysteric request from the public driven by the Oil-Shock in 1972, and also because of the scientific and political competition between nations during the Cold War. These unusual conditions helped to make the laser systems bigger and bigger.

I have seen and experienced such things in my 25 years career as fusion-related scientist. In this Teller Medal lecture, I summarize my main contributions to inertial fusion energy (IFE) and high-energy-density (HED) physics. In addition, I mention the future direction I intend to go based on the discipline cultivated in IFE. I would like to report five topics carried out in such historical background.

2. Anomalous Transport of High Energy Electrons

First one is based on my Ph. D thesis at Osaka University. At that time, flux limitation of electron heat transport was one of the hottest topics and Dr. K. Mima mentioned me that nonlinear plasma physics is important and may cause the anomalous transport through ion wave turbulence. He suggested me as follows. The penetration of the hot electrons induces the return current so as to keep charge neutrality. The current, however, makes the ion wave unstable through the inverse-Landau damping. In order to estimate the effective resistivity to the return current, I have to calculate the saturated amplitude and spectrum of the ion wave turbulence.

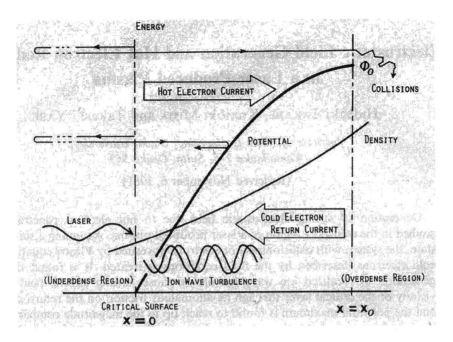

Figure-1: *Schematics of anomalous transport to the hot electrons generated in the critical density region. The penetration of the hot electron current induces the return current by cold bulk electrons. Then, the return current induces ion wave instability through the inverse-Landau damping. The ion wave finally becomes turbulent and the scattering of the return current by the turbulent fields induces anomalous resistivity. This enhanced resistivity increases the height of electrostatic fields following Ohmic law. The height of the electrostatic field is found to be almost same as the average kinetic energy of the hot electrons. The electrostatic field, as the result, inhibits the penetration of high energy electron and this is the mechanism of anomalous transport of hot electrons. Since the am-bipolar field is generated in the expansion region, relatively low energy hot electron component is confined and only the high energy component can penetrate into over-dense region.*

The saturated amplitude and spectrum of the ion wave turbulence are calculated with the renormalization theory according to a book by Kadmtsev [1]. I found that almost the same value of the potential hill is generated as the average kinetic energy of the hot electrons[2]. These physics scenario is shown schematically in Fig. 1. After this work, I found that in the same situation, not only such two-stream type instability, but also Weibel-type instability with growth of magnetic field becomes important, and the penetrating hot electron current is broken up into many filaments. Recently I came to the following understanding that my Ph. D work is strongly related to the anomalous transport of relativistic electron energy flow in Fast Ignition scheme. I will discuss more details on this point in Section 7.

3. Rayleigh-Taylor Instability at Ablation Front

After finishing my Ph. D thesis, I worked for Max-Planck Institute for Plasma Physics as guest researcher with Peter Mulser in Laser Research Group, which soon after became Max-Planck Institute for Quantum Optics [3]. I worked on a nonlinear physics in laser plasma interactions and obtained a self-consistent plasma wave profile in sub- to super-sonic transition flow driven by the plasma wave ponderomotive force[4].

I moved to USA to work in the group of Prof. R. L. Morse at University of Arizona, Tucson. During the stay in Tucson, I could focus my attention on Rayleigh-Taylor instability at the ablation front. This work was in progress by L. Monierth, time-then a graduate student, by solving equations to linear perturbations with finite difference method in order to get the dispersion relations for a variety of ablation structure. In addition, it became later clear that the growth rate at shorter wavelength is not reliable. I proposed to solve the problem as 5th order eigen value problem. When solving the linearized equations as eigen value problem with combination of numerical and analytical mode structures, I could improve the numerical scheme so that extremely high accuracy is guaranteed by controlling steps for integration. This numerical method to solve the fifth order eigen value problem was itself original and was published in Ref. [5].

When I returned to Osaka, I summarized all data for the dispersion relation in normalized form and obtained the following simplified formula:

$$\gamma = \alpha(kg)^{1/2} - \beta k v_a \tag{1}$$

where, v_a is the ablation velocity at the ablation front and α and β are factors numerically fitted to the eigen-value data[6]. What was surprised for me was that the stabilization given in the term β is much larger that unity suggested in Ref. [7].

It is my great honor that Eq. (1) is now called *Takabe formula* worldwide. The number of citations in ISI database is 87 for Ref. [5] and 177 for Ref. [6] by the mid of 2003. After several years from the publications, the formula has been verified at first with two-dimensional codes in USA. Tabak et al have compared systematically the formula with the growth rate obtained with LASNEX code and concluded that the formula well fits the LASNEX result [8]. Soon after, Gardner et al at Naval Research Laboratory have published their simulation results with FAST-2D and insisted a good coincidence with the formula [9]. The same discussion was also seen in Report of LLE, Univ. Rochester with OHCHID code [10].

The formula has also been checked afterward experimentally in the indirect scheme and direct scheme. The former has been done at LLNL by B. Remington et al [11]. In the case of indirect scheme, the radiation uniformity is easily kept and experiment is rather easy to be done. In the case of the direct scheme, a series of experiments have been done [12], after smoothing of beam profiles. Most recent status was presented by H. Shiraga in this conference[13].

4. Integrated Code

After completion of the study of Rayleigh-Taylor growth at ablation front, I was involved in the implosion experiment at Osaka University. Initially, I tried to solve analytically the implosion dynamics and instability issues [14]. Even with an elegant solution based on a self-similar solution and its stability analysis, however, this was not the final answer. After reading the book "Enrico Fermi" by Emilio Segre [15] I was encouraged to start development of a new implosion code. I had studied mathematics of finite difference method at Arizona and could not feel any difficulty to develop the program of compressible hydrodynamics.

Initially, the code is of course a simple one-dimensional one. Once the Gekko XII laser was completed and a series of implosion experiments have started, I was required and expected not only to analyze the experimental data, but also to design targets for experiments. I have to make my code integrated physics one. It is, however, very pleasant job for me. Laser coupling was installed, Non-LTE atomic model based on AAM(average atom model) was installed, radiation transport kinetics with multi-group flux-limited model was installed, fitting formula for equation of state based on Thomas-Fermi model has been obtained to be installed in the code, and so on. The report on the physical and mathematical modeling of the code, which was named ILESTA, has been written in English[16], while it is my regret that I lost a chance to publish it.

It should be noted that AAM even present-day used is originally studied by supervision of E. Teller[17].

One of the interests to develop an integrated code for laser implosion was to find the relation to astrophysics in the published literature showing the similarity to the physics of inertial confinement fusion. When laser fusion research began, the textbooks of astrophysics gave us guidance on

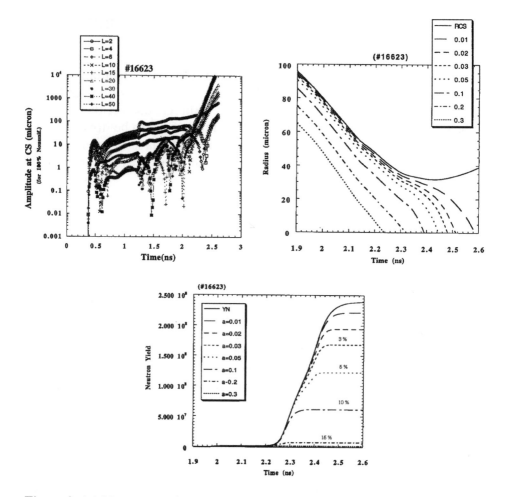

Figure-2: *(a) Linear perturbation growth for l = 2 – 50 calculated with ILESTA-2D code. In the final stagnation phase (t = 2-2.5 ns), shorter wavelength modes growth explosively. (b) The trajectories of effective mixing zone. The solid line is the trajectories of 1-D simulation. (c) The effective neutron yields calculated as neutrons produced inside the mixing zone front of Fig. (b).*

radiation transport, equation of state, opacity and so on. Among them, theoretical papers dealing with high-density plasmas can be found only related to astrophysics and I remember I studied many of them when I was a student.

During the campaign of highest neutron yield with high-aspect-ratio glass micro-balloons, ILESTA code has been used to analyze experimental data [18], reveal scaling law of the implosion performance [19], and so on. It is almost this time that ILESTA code was upgraded to two-dimensional one [20].

Throughout the implosion experiment, it was found that the turbulent mixing became the most critical physics to degrade the neutron yield, and I tried to install the turbulent mixing model in ILESTA code. By treating the mixing in the final stagnation phase, we could reproduce the experimental data[21]. This mixing model is so-called k-ε type turbulence model and additionally we have to solve time evolution of turbulence energy k and its dissipation rate ε. The effect of turbulent mixing at the ablation front in the acceleration phase has also been studied analytically based on quasi-linear theory[22]. This gave us the critical condition of the survival of accelerated shell so that the in-flight aspect ratio should be less than 70.

After the campaign of beam smoothing of Gekko XII laser, we have carried out the implosion experiment of plastic shell with DD gas [23]. With the 2-D version of ILESTA code, I studied temporal growth of the linear perturbations and effective implosion performance was evaluated based on Haan's mix model[24]. In Fig. 2(a), the time evolution of the perturbation amplitude at the contact surface is plotted for the mode $l = 2$ ~ 50. The perturbations are induced by laser non-uniformity and the vertical scale corresponds to the displacement of the contact surface in μm unit for the case of 100% laser non-uniformity. It is clear that the stagnation is very unstable and the perturbation grows abruptly as seen in Fig. 2(a). In Fig. 2(b), the trajectory of the contact surface is plotted around the final compression. The solid line(RCS) is the case of 1-D simulation without mixing. The effective mixing front calculated by the sum of all perturbations with Haan's saturation model is plotted in Fig. 2(b), where each line represents the case with the fraction of the total laser non-uniformity RMS from 0.01 to 0.5. There have been published about laser nonuniformity spectrum[25], indicating enhanced amplitude in lower l modes. In this calculation, however, we assumed the white noise of irradiation non-uniformity spectrum, because the lower modes stem from multi-beam overlapping and its contribution was not quantitatively evaluated at that time. In Fig. 2(c), the time-integrated effective neutron yield defined to be the total neutrons inside the mixing front is plotted. In order to explain the experimental data, it is found that we need to assume the laser non-uniformity about 15 %(RMS), which roughly coincides with the laser data.

In order to confirm how non-uniform the implosion dynamics is, I finally carried out axially symmetric two dimensional simulations with surface tracking technique[26]. The typical example at the maximum compression is shown in Fig. 3. In Fig. 3, the DD fuel concentration and temperature profiles are shown at the center, and contours of the density and temperature are plotted on the both sides. In this 2-D simulation, we

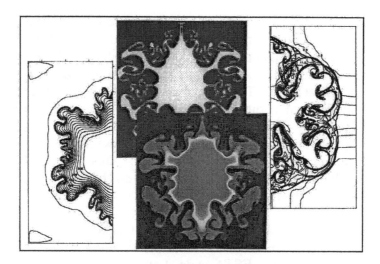

Figure-3: *Snap shots of the implosion same as in Fig. 2. The time corresponds to that at the maximum neutron generation rate. At the center, the DD fuel concentration and temperature distribution are shown. At the right, the density profile is shown in contour, while at the left the temperature is shown in contour. With this two dimensional simulation we could obtain the same number of the total neutron yield as experiment. The increase of the effective area of the fuel-pusher contact surface enhances the cooling by electron thermal conduction.*

have obtained almost the same amount of neutron yield. From the DD concentration profile, it is clear that the central part is compressed, while the outer region goes out with finger structure. In the density contours, the mushroom structures are seen in the relatively low temperature region, while the mushroom is not seen for the fingering high-density parts penetrating into high temperature region. This is clear evidence of the fire-polishing due to the thermal conduction.

More detail analysis with 2D code coupling with spectroscopic post-processor has been done and comprehensive analysis is given in Ref [27].

5. Laboratory Astrophysics -*Historical Background*-

When the short pulse and intense lasers came into being familiar and the word "laser-plasma" was born, John Dawson published a paper on the properties and application of the laser-plasma[28]. In that paper there is a suggestion that intense lasers can be used as a simulator of astrophysics and that experiments related to the collision-less shocks generated by the explosion of a supernova or solar-flare phenomena can be studied in laboratory. This paper, however, did not trigger astrophysics experiments. At that time nobody knew what sort of plasma the laser-plasma is, and it was enough work to generate, diagnose, and analyze the laser-plasma. In addition, when matters are irradiated by lasers, an anomalous absorption was measured and a variety of nonlinear phenomena such as the parametric instabilities were observed. Therefore, it was not the time to apply the lasers to fusion or study of astrophysics. First it was necessary to understand the physics of laser-plasma.

After the Oil-Shock in 1972, the construction of big laser systems made it possible to start to understand implosion physics and laser fusion energy research progressed dramatically. This was the time of 1980's. Around that time, B. Ripin and J. Grun et al. in Naval Research Laboratory carried out a series of experiments on blast waves generated by the laser system "Pharos"[29]. By taking detailed data with an interference technique, they checked the self-similarity of the wave front propagation and observed the hydrodynamic instability of the blast wave front. The time evolution of the turbulent spectrum in the later stage of the instability was measured. In addition, they also carried out an experiment on a blast wave in a gas with an externally imposed magnetic field and discussed the phenomena of reconnection of the magnetic field [30]. The same sort of experiment was also carried out in the former Soviet Union, and they also mentioned that the experiment has been done to study the interaction between the blast wave generated by a supernova explosion and the inter-stellar magnetic field [31]. There were, however, no continuing activities after these publications, although they titled their activity as science for astrophysics. One of the reasons is that their work was related to the SDI (Strategic Defense Initiative) program and the magnetic field modeled the earth's magnetic field to see the behavior of expanding plasma in the earth's magnetic field when a nuclear weapon is exploded. This was a reasonable activity at that time from the view point of the government which funded the SDI program, for example, in US.

At almost the same time, S. J. Rose[32] and R. W. Lee[33] have published proposals regarding astrophysics related laser plasma experiments. They mainly emphasized that the laser plasma can be used to study the radiative properties of stellar media, for example, opacity, equation of state, radiation heating and so on. It is noted that their proposal is accepted by many groups as opacity experiment in laboratory, and opacity workshops have been held routinely. This is also one side of Laboratory Astrophysics, mainly focusing "sameness" of the plasma I will discuss later. It is a good counter part compared to that NRL scientists mainly focused on "similarity" of dynamics of astrophysical phenomena.

On the other hand, the supernova 1987A (SN1987A) observed in the south hemisphere on February 23, 1987 taught us that hydrodynamic instability is a big issue not only in the field of laser fusion research [34]. The explosions of supernovae are among the most spectacular physical events in the Universe. Elements heavier than helium are synthesized in massive stars and finally scattered out into space through the supernova explosion. Such ejected materials became the source to form the earth after billions of years. This story encourages us to study physics. To analyze the X-ray and Gamma-ray data from SN1987A the possibility of hydrodynamic instability and mixing in the explosion has been intensively discussed. This spectacular fact of a deep relation of hydrodynamic instability to physics of supernova explosion taught us that the laser fusion scientist is working for not just physics in 1 mm space, but revealing the deep inside of the nature seen in much broader fields. This encouraged us and taught us the importance of having widely opened eyes.

This is the start for me to get heavily involved in almost all field of astrophysics. And it was a starting point of a new field in astrophysics and HED physics. It is now called *Laboratory Astrophysics* or *Laser Astrophysics* in short. It should be, of course, noted that there is institutes with this words, for example, Kellogg radiation laboratory at Cal Tech and Joint Institute for Laboratory Astrophysics (JILA) in University of Colorado, Boulder. JILA is a very famous institute, while the difference is that our new Laboratory Astrophysics (LA) deals with model experiments of astrophysical phenomena, namely reproducing a variety of dynamical phenomena. The main interest of JILA has been focused on atomic and molecular physics and database. Laboratory Astrophysics also includes the study of astrophysics related physics with accelerators. For example, the structure and properties of unstable nuclei are intensively studied recently in relation to the nuclear database for R-process important for nucleosynthesis in gravitationally collapsing supernovae. It should be noted that almost all laboratory astrophysics studies with accelerator are to obtain database and any model experiments have not been done to study the collective dynamics of astrophysical phenomena, except one case which is QGP(quark-gluon-plasma) experiment with RHIC at Brookheven National Lab[35].

Anyhow, throughout this explosion of SN1987A, we had two big discoveries. This is well shown in the drawing by the late Prof. Minoru Oda, the pioneer of X-ray astrophysics in Japan, given in Fig. 4[36]. One is the first detection of astronomical neutrino by Kamiokande, 3000 ton pure water tank located deep in the old mine tunnel in Japan. The observed eleven neutrinos were the evidence of validating the theory of gravitational collapse of a massive star and this observation had led Prof. Koshiba to Nobel Prize in Physics, 2002.

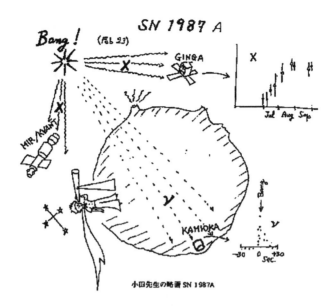

Figure-4: *An image of the explosion of SN1987A drawn by late professor Minoru Oda, pioneer of X-ray astronomy in Japan. Eleven neutrinos from the core-collapsing supernova has been observed with KAMIOKANDE, and the theory of gravitational collapse has been validated. The early appearance of Comptonized hard X-ray was observed GINGA x-ray satellite, and hydrodynamic instability was found to be important in the explosion.*

The other big discovery came from Japanese X-ray satellite GINGA. The satellite has observed hard X-ray in the range of 10 – 20 keV from SN1987A. The X-ray stems from the γ-ray generated by nuclear decay deep inside of the star. The Compton scatterings make the γ-ray soft and it diffuses out to be observed as hard X-ray. According to one-dimensional simulation of supernova explosion, the timing of the appearance of such X-ray has been predicted to be almost one year later, because it takes time so that the column density of the star become low and the star is transparent to such X-ray generated inside. By the way, such success of the X-ray astronomy gave Nobel Prize to Prof. Giaconi also in 2002.

GINGA satellite started its mission just before the explosion and aimed at this X-ray from SN1987A. It was surprising that it observed this X-ray only after three month and proofed that simple one-dimensional simulation can not explain what is going on inside SN1987A. Prof. Ken Nomoto, University of Tokyo, is Supernova theoretician and faced on a difficulty to explain quantitatively. It was clear that hydrodynamic instability and consequent mixing process may explain the observation data, while it is not able to be simulated at that time. He called me and suggested me to join his group to challenge new field of multi-dimensional simulation of supernova explosion. This is the trigger for me to get into astrophysics deeply. It should be noted that as you see in Fig. 4, γ -ray satellite of former Soviet Union, KVANT, also supported the mixing inside SN1987A. I recommend reader interested in the science of SN1987A to read review papers [34].

After several years work, my contributions were accepted by the astrophysicists in Japan. In many conversations with them, what I found is that they have no confidence of their simulation codes because there is no way to check the codes by comparing with any experiments. Of course, it is difficult to do any model experiments to check high-temperature, high-density physics in laboratory. Through such conversations, I proposed to use lasers to do model experiments of astrophysical phenomena and use the data to check the validity of their computer codes.

Hydrodynamic instability of supernova explosion was my starting point of this new academic field of Laboratory Astrophysics. Bruce Remington at LLNL was the first experimentalist who could understand the importance of this new field and he started model experiments mainly regarding hydrodynamics with NOVA laser at LLNL [37]. With encouragement from both sides of astrophysicists and laser experimentalists, we have carried out a case study on what kind of model experiments we can do [38]. In Table-1, I summarized 18 model experiments of astrophysics in the crossover of six physics issues in laser fusion and three categories of the way of thinking. I briefly explain three of them.

6. Laboratory Astrophysics -*Three Views and Three Purposes-*

An example to explain the first purpose to promote LA is No.13 in Table-1, Vishiniac instability of blast wave. It is very interesting that Grun et al found a turbulent blast wave when they produced a blast wave in xenon gas [39, 29]. In case of nitrogen gas, however, they observed a clear spherical blast wave. The difference is concluded due to radiation cooling effect. Radiation cooling is enhanced in xenon gas and the density jump at the blast wave front increases. Since the blast wave front is decelerated and is unstable to Rayleigh-Taylor type instability, although the super sonic flow across the discontinuous surface help the front stable. In fact, once the surface bends, the shock condition across the surface makes the flow so that the pressure behind the dip increases. As the result, the blast wave front becomes over-stable and perturbation of the blast wave front grows with oscillation as pointed out by Vishinac [40].

After this experiment, astrophysicists had studied analytically and numerically if the same instability may occur in astrophysical blast waves [41].

	A. Sameness	B. Similarity	C. Resemblance
1. Laser Plasma Interactions			
2. Electron Energy Transport			1. Non-local Transport of Neutrino in Baby Neutron Star
3. Hydrodynamics and Shocks	2. EOS Giant Planets	3. Interaction of Molecular Cloud with Strong Shock by SNR (Morphology)	4. Collision-less Socks and Particle Acceleration (Origin of Cosmic Ray)
4. Hydrodynamic Instabilities		5. RT Instability of SN Explosion	6. Hydro-instability in Neutrino driven SN Explosion 7. RT Instability of Eagle Nebula
5. AtomicPhysics and X-ray Transport	8. Opacity Evoluti on of Stars etc.	9. Atomic Process in SNRs 10. Stellar Jets 11. Radiation Hydro. in Early Galaxy 12. Photo-ionized Plasmas	13. Vishiniac Instability of SNRs 14. X-ray Laser in Universe
6. Laser-produced Relativistic Plasmas			15. Fireball of Gamma-ray Bursts 16. Cosmological Jets (Rel.) 17. Weibel Instability in GRB 18. Non-LTE QED at AGN Core

Table-1: *Crossover of laser fusion and laboratory astrophysics. By picking up physics subject of laser fusion with viewpoints of sameness, similarity, and resemblance, we can enumerate at least 18 model experiments for Laboratory astrophysics at the present time.*

Recently, US X-ray satellite Chandra observed a striking image of the Tycho SNR (Supernova Remnant) as shown in Fig. 5(a) [42]. The x-ray image shows the turbulent debris created by a supernova explosion that was observed by the Danish astronomer Tycho Brahe in the year 1572. The shadow image of laser produced turbulent blast wave is also shown in Fig. 5(b) for comparison. The spatial scale is of more than 18 orders of magnitude different. We can see that the turbulent blast wave has been

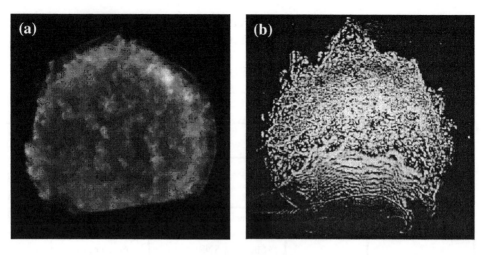

Figure-5: *Turbulent blast waves in Universe (a) and in laboratory (b).*
Figure (a) is X-ray image of Tycho SNR by Chandra X-ray satellite showing the turbulent debris, while Fig. (b) is the shadow image of laser produced blast wave which became turbulent due to hydrodynamic instability. The spatial size is more than 18 orders different.

predicted with laboratory experiment amore than 10 years ago. Such discovery of new phenomena through laser model experiments is one of three main purposes why we promote Laboratory Astrophysics. We call such discovery "serendipity", the importance of which has been proofed in the history of science and we can enumerate many examples.

The second purpose of LA is to do model experiment for verification and validation of numerical code used in astrophysics. For this purpose, B. Remington et al [37] has carried out a variety of experiment and the resultant data has been compared with a typical code for supernova explosion PROMETHEUS. This subject is No. 5 in Table-1. More comprehensive and systematic study for validation of astrophysical code FLASH [43] has been carried out by focusing on two cases[44]. One is laser-driven shock passage through a multi-layer target, a configuration subject to both Rayleigh-Taylor and Richtmyer-Meshkov instabilities. The second test is the classical Rayleigh-Taylor instability. They obtained a good agreement in the former, while not in the latter case, and discussed new findings through the comparison. I recommend readers who interested in a variety of experiments in this category to refer Ref. [45].

The third purpose is well described, for example, related to the subject 17 in Table-1. We aim to export mature knowledge of physics accomplished in laser plasma field to astrophysics. The Weibel instability and related

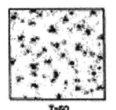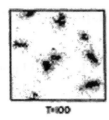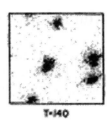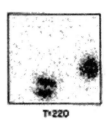

Figure-6: *Two dimensional simulations of nonlinear evolutions of Weibel instability. Normalized time is 50, 100, 140, and 220 from left to right. Uniform electron beam is assumed to propagate perpendicular to the paper. The beam becomes unstable and split into many filaments, because the linear growth rate of Weibel instability is large for shorter wavelength. Because of the magnetic force surrounding the filaments induces the coalescence of filaments; namely, magnetic reconnection occurs and beam energy is dissipated anomalously. As the result, larger scale structures are produced and total energy propagating as beam reduced drastically.*

structure formation of magnetic field by relativistic electrons have been studied intensively in relation to Fast Ignition[46]. Even before Fast Ignition research, this was an important subject in REB fusion research[47], where REB means relativistic electron beam. What we found was that the shorter wavelength magnetic perturbation grows predominantly in linear instability phase, while small scale current filaments coalescence each other to form a large scale magnetic field as seen in Fig. 6. Then, the electron energies are spent for heating background plasma through the electric field generated by the reconnection of magnetic field surrounding the filaments. As the result, anomalous resistivity appears in propagation of relativistic and intense electron current. The most fascinating subject in astrophysics is now Gamma-ray Bursts (GRB)[48]. It is reported in the summer 2003 that the γ-ray stemming from the synchrotron emission in the super-relativistic electron-positron plasmas is lineally polarized by about 80%[49]. This indicates that the structured magnetic field exits in the super-relativistic jet. I believe that the structure formation scenario in laser plasma is also seen in the super-relativistic jet.

It is useful to summarize the importance of these three viewpoints in promoting Laboratory Astrophysics.

(1) New finding (serendipity) can be expected in laser model experiments.
(2) Model experiments can be used to validate and verify the simulation codes to be used for astrophysics research.
(3) Mature science accomplished in laser plasma can be exported to understand mysterious physics in Universe.

7. Comment on Anomalous Transport in Fast Ignition

I would like to point out the nonlinear phenomena of Weibel instability is essential subject for Fast Ignition study. It is well known that the Alfven current limits the total current propagating in vacuum and it is written to be

$$J_A = 17\beta\gamma \ [kA] \tag{2}$$

where β =v/c and γ the Lorentz factor of the beam with velocity v. Then the power per beam we can transport is limited to be

$$P_A = 9\gamma^2 \ [GW] \tag{3}$$
$$(\ = 1 \ TW \ for \ \gamma = 10 \ ).$$

This means, we have to keep about 10,000 filaments remain so that the order of 10 PW energy penetrates into over-dense region to heat and form ignition region in a compressed core[50]. This scenario, however, is found not to work with the last ten years studies mainly with PIC code.

The scenario of the linear growth of Weibel instability, formation of many current filaments, coalescence of filaments, and large-scale structure formation of magnetic field is estimated to happen in the time scale of sub-pico-second. In Fast Ignition, ignition demonstration requires transport of several PW of relativistic electron energy over several pico-seconds. There is no theory and no simulation that seriously studied such long time evolution self-consistently. I would like to predict here what would happen in such long time evolution.

Plausible physics scenario is as follows:

1. Weibel instability growth and current filament formation
2. Saturation of instability and filament coalescence
3. Magnetic field reconnection dissipates electron energy of filaments and heats up the background plasma
4. Weibel instability becomes stabilized because of the increase of the plasma temperature
5. Counter-streaming return current is produced
6. Ion waves are induced by the return current as explained in Section 2.
7. Ion wave turbulence causes anomalous resistivity to penetration of high energy electrons and the resultant stopping distance becomes anomalously short compared to that estimated classically.

I guess in Fast Ignition scheme, the phenomena 1 – 4 are seen in sub-pico-second range and 5 –7 may be seen in pico-second range. Or, the phenomena 1-7 may repeat like relaxation oscillation over relatively long time. This means that the key physics if Fast Ignition does work or not calls the revival of my PhD work in higher density region, where the ion plasma frequency is $\omega_{pi} \sim 1$ fs^{-1}. I hope this time my former supervisor and the leader of Fast Ignition, Prof. Mima, gives me answer to this very attractive physical problem according his guideline on the importance of anomalous effect in plasma physics.

8. Computational Observatory

In order to promote Laboratory Astrophysics, two activities are essential.

1. Model experiment in laboratory, namely LA explained in Section 7.
2. Development of integrated codes of astrophysics to be validated through comparison.

It is a grand challenge for me to extend my career to developing an integrated code of astrophysics systematically.

In case of gravitationally collapsing supernova like SN1987A, the key physics is transport of neutrino and multi-dimensional effect of hydrodynamics. The shock wave produced at first by the bounce of the collapse of iron core is said to disappear because of energy absorbing nuclear reaction should be re-birthed by neutrino heating. As recently reported[51], however, it is very hard to obtain the result with explosion because the neutrino heating is not enough to keep the strength of the shock wave so as to explode the surrounding materials. The physics of non-local neutrino transport is essential as non-local electron transport is important in direct drive laser fusion. This is listed as No. 1 in Table-1.

Supernova explosion is the world of integrated physics. Multi-dimensional hydrodynamics should be solved by coupling with gravitational force, nucleosynthesis, radiation and neutrino transport, relativity effect, and so on. Such grand challenge is very exciting. I am collaborating with 15 groups over Japan to promote the Computational Observatory project as reported in Ref. [52].

The origin of heavy elements stems from the nucleosynthesis in massive stars and supernova explosions. I have carried out calculation of application of lasers to nuclear transmutation through giant dipole resonance of nuclei [53]. Ultra-intense lasers have a potential ability to be study nuclear physics and, for example, the structure and physics constants of unstable nuclei which become essential in R-process nucleosynthesis in supernova explosion could be studied with use of lasers.

9. Summary

I have summarized a part of my activities over my 25 years career in laser plasma, laser fusion, and related scientific fields.   I have briefly explained five topics I was involved and I now work.   The main concern has shifted from implosion and fusion research to fundamental science by mainly focusing Laboratory Astrophysics.   I here summarized this paper.

(1) It is concluded that the anomalous transport driven by ion wave turbulence induced by return current to compensate charge neutrality is essential in laser plasma in order to treat hot electron transport. This 20 year-old work revives as key issue of physics for studying the plasma physics of ultra intense current penetration in the compressed core in Fast Ignition scheme.

(2) Enhanced ablative stabilization was predicted and validated.   It is the core physics of ICF and no hope of ignition and high-gain without this ablative stabilization mechanism.

(3) Through Integrated Code development for ICF, not only I could contribute to ICF, but I also found the high-energy-density physics is very attractive.

(4) Laboratory Astrophysics has itself a long history and became very visible recently.   It is important field for Initial Fusion Science to attract excellent students and scientists.

(5) I am now promoting Computational Observatory Project to build more sophisticated integrated code of astrophysics. Such inter-disciplinary activity will make our society of Inertial Fusion Science more mature and make high-energy-density physics very attractive for all scientists.

In US, NIF(National Ignition Facility) is under construction[54].   In the web-site and the presentation by Dr. G. Miller in this conference, three items are refereed as the mission of NIF project.   They are "national security", "basic science", and "energy".   They were four items including "industrial competitiveness" in an old report[55].   The scientific research always has two faces and this tendency was enhanced during the Cold War.

The Cold War was over and a new world political system is going to be constructed. However, the world contains still many elements of un-stability and uncertainty. In such situation, it is not so easy to rapidly change the direction and strategy of research. With the end of the Cold War, LLNL seems to reexamine its future plan in order to become the first runner of the world in the field of fundamental science [56]. It is, however, unavoidable to regard that the first priority of NIF is the national security at the present time. Through the present era of the globalization of economics and politics and rapid change of social system, it is not sure that the above order of the three items remains the same until the completion of the NIF. As the accelerator has grown after World War II as a big device for fundamental science to study elementary particle physics and nuclear physics, this time the laser has a potentiality to be regarded as a big device to study a new fundamental science in the near future. In this course, the relation with astrophysics, which is an everlasting academic subject for the human beings, is very important.

The computer is a well-known example of the dual use of science. Physics Today journal published a special issue titled "50 years of computers and physics" almost eight years ago [57]. In the issue, the cradle of the scientific computer was described in the article titled "From Mars to Minerva: The origins of Scientific computing in the AEC Labs" (by R. W. Sedel). It is the fact that the origin of computer is "Mars"(God of War). It is clear evidence of the wisdom of the human beings to have changed it to "Minerva" (God of Intelligence). The same should be done concerning all science and technology, I think.

The astrophysics with intense laser is not only a new trend of research, but also should offer students and young researchers a good opportunity to satisfy their scientific curiosity. If the astrophysics with lasers grows as one academic field, I believe that it will provide a good place to educate experts for long-term fusion energy research.

Finally, I hope the works I have been concerned and mainly related to laser fusion will purely grow from the discipline of Mars to the discipline of Minerva in the near future.

Acknowledgments

The author would like to thank his teachers, colleagues, friends, and students. He especially thanks Prof. Chiyoe Yamanaka and Prof. Kunioki Mima for their guidance and encouragement. He would like to thank Prof. Peter Mulser and Dr. S. Witkowski for their hospitality during his stay in Munich.   He would like to thank Prof. Richard L. Morse, Dr. Patrick McKenty, and Dr. Leland Montierth for their kindness at University of Arizona, Tucson.   He thanks Prof. Ken Nomoto for his guidance of the author's career to astrophysics.   He also thanks Dr. Bruce Remington and Dr. Mike Campbell for encouraging him to promote Laboratory Astrophysics.   Finally, he would like to say many thanks to his students, who have been my metal supporters and co-explorers in proceeding towards new subjects.

References

[1]   B. B. Kadmtsev, *Plasma Turbulence*, (Academic Press, London, 1965).
[2]   H. Takabe et al., J. Phys. Soc. Japan **51**, 2293 (1982).
[3]   MPQ web-site: http://www.mpq.mpg.de/
[4]   H. Takabe and P. Mulser, Phys. Fluids **25**, 2304 (1982).
[5]   H. Takabe, L. Montierth, and R. L. Morse, Phys. Fluids **26**, 2299 (1983).
[6]   H. Takabe, et al., Phys. Fluids **28**, 3676 (1985).
[7]   S. E. Bodner, Phys. Rev. Lett. **33**, 761 (1974).
[8]   M. Tabak et al, Phys. Fluids **B2**, 1007 (1990).
[9]   J. Gardner et al., Phys. Fluids **B3**, 1070 (1991).
[10] LLE Review, LLE, Univ. Rochester **37**, 2 (1988).
[11] B. Remington et al, Phys. Plasmas **2**, 241 (1995).
[12] K. Shigemori et al., Phys. Rev. Lett. **78**, 250 (1997); S. G. Glendinning et al., Phys. Rev. Lett.**78**, 331 (1997); J. P. Knauer et al., Phys. Plasmas **7**, 338 (2000).
[13] H. Shiraga, WP3.4, Plenary at IFSA03, in this Proceedings.
[14] For example; F. Hattori, H. Takabe, and K. Mima, Phys. Fluids **29**, 1719 (1986).
[15] E. Segre, *Enrico Fermi Physicist*, (Univ. Chicago Press, 1970).
[16] H. Takabe, *Hydrodynamic Simulation Code for Laser Driven Implosion - Physics and Algorithm of ILESTA Code - ,* (1994) unpublished.
[17] H. Mayer, *Method of Opacity Calculations*, Los Alamos Report LA-647 (1947) in Introduction.
[18] H. Azechi et al., Appl. Phys. Lett. **55**, 945 (1989).
[19] H. Takabe et al., Phys. Fluids **31**, 2884 (1988).
[20] H. Takabe et al., Laser and Particle Beams **7**, 175 (1989); H. Takabe and T. Ishii, Jpn. J. Appl. Phys. **32**, 5675 (1993).
[21] Y. Isayama and H. Takabe, J. Plasma Fusion Res. **70**, 756 (1994) *in Japanese*.
[22] H. Takabe and A. Yamamoto, Phys. Rev. **A44**, 5142 (1991).

[23] K. Mima et al., Phys. Plasmas **3**, 2077-2083(1996).

[24] S. Haan, Phys. Rev. **A 39**, 5812 (1989).

[25] S. G. Glendinning et al., Phys. Rev. **E 54**, 4473 (1996); D. H. Kalantar et al., Phys. Rev. Lett., **76**, 3574 (1996).

[26] H. Takeuchi, Bachelor Thesis, Osaka University 1995 *in Japanese*.

[27] H. Takabe et al, Plasma Physics & Control Fusion **41**, A75-98 (1999).

[28] J. M. Dawson, Phys. Fluids **7**, 981 (1964).

[29] B. H. Ripin et al, Laser & Part. Beams **8**, 183 (1990); J. Grun et al, "Experimental Studies of Very High Mach Number Hydrodynamics", NRL Mem. Rep. 6790-94-7366, 1994.

[30] B. H. Ripin et al., Phys. Fluids **B5**, 3491 (1993).

[31] V. M. Antonov et al, "Laser Interaction with Matter", ed. S. Rose, IOP Conf. Ser. No. 140 (IOP, 1995), p. 167.

[32] S. J. Rose, Laser & Part. Beams **9**, 869 (1991).

[33] Web site; http://www.llnl.gov/science_on_lasers/ or R. W. Lee, "Science on the NIF" in "Energy & Technology Review", LLNL, December 1994, pp. 43-54.

[34] For example; W. D. Arnnet et al., "Supernova 1987A", Annu. Rev. Astron. Astrophys. 1989, **27**: 627-700.; W. Hillebrandt and P. Hoflich, "The Supernova 1987A in the Large Magellanic Cloud", Rep. Prog. Phys. **52**, 1421-73 (1989).; H. Takabe, "Inertial Confinement Fusion and Supernova Explosion", Japanese J. Plasmas & Fusion Res., **69**, 1285 (1993) *in Japanese*.

[35] http://www.bnl.gov/rhic/

[36] Y. Tanaka, *Obituary: Minoru Oda* (1923-2001),Tenmon-Geppo (2001) *in Japanese*.

[37] B. A. Remington et al., Phys. Plasmas **4**, 1994 (1997); B. A. Remington et al, Science **284**, 1488 (1999).

[38] H. Takabe, Progress of Theoretical Physics Supplement, No. **143**, 202-265 (2001).

[39] J. Grun et al., Phys. Rev. Lett. **66**, 2738 (1991)

[40] E. T. Vishniac and D. Ryu, Astrophys. J. **337**, 917 (1989).

[41] M. M. Mac Low and M. L. Norman, Astrophysical J. **407**, 207 (1993).

[42] http://chandra.harvard.edu/

[43] http://flash.uchicago.edu/

[44] A. Calder et al., Astrophys. J Suppl. **143**, 201 (2002).

[45] Edited by B. Remington, et al., Astrophysical Journal; Supplement **127**, No2, Part 1, (2000).

[46] W Y. Sentoku et al., Phys. Plasmas **7**, 689 (2000)

[47] R. Lee and M. Lampe, Phys. Rev. Lett., **31**, 1390 (1973).

[48] N. Gehrels et al., Scientific American, December (2002).

[49] W. Coburn and S. E. Boggs, Nature **423**, 415 (2003).

[50] M. Honda, J. Meyer-ter-Vehn, and A. Pukov, Phys. Plasmas **7**, 1302 (2000)

[51] R. Buras et al, Phys. Rev. Lett. **90**, 241101 (2003)

[52] H. Takabe, *Computational Observatory*, J. Plasma Fusion Res. **79**, 504 (2003) *in Japanese.*

[53] H. Takabe, *Laser Nuclear Physics*, AAPPS Bulletin, Vol.**13** No.1 (2003), pp. 18-25, Download at http://www.aapps.org/.

[54] NIF web-site; http://www.llnl.gov/nif/

[55] ICF Quarterly Report, Vol. 7, No.3 (April-June, 1997), Report is available with Web-site; (http://lasers.llnl.gov/lasers/pubs/icfq.html)

[56] Science **275**, 1252 (1997).

[57] Physics Today, October issue, 1996.

30 YEARS LASER INTERACTION AND RELATED PLASMA PHENOMENA
(a reflection by H. Hora at the 13th conference, Monterey 1997)

30 years LASER INTERACTION AND RELATED PLASMA PHENOMENA*

Heinrich Hora

Department of Theoretical Physics, University of New South Wales, Sydney 2052, Australia

When these proceedings of 13th international conference LASER INTERACTION AND RELATED PLASMA PHENOMENA (LIRPP) will be circulated in 1998, it is just 30 years that this conference series began. Professor Miley asked me to present some thoughts at this occasion since I am involved from the beginning to 1991 a director and then as emeritus director. The conferences were in the following years 1969, 1971, 1973, 1976, 1979, 1982, 1985, 1987, 1989, 1991, 1993, 1995 and 1997 and reference to each of the conferences is simply given by the year in brackets.

The conference began with an interview I had May 1968 by Warren Stoker, the Vice President of the Rensselaer Polytechnic Institute, Graduate Center in Hartford, Conn.. Good University Presidents have the unique ability to induce exciting ideas and proposals in others to whom they are talking. So I was envisaging for his institute to found an international conference on the problems of high intensity lasers and the interaction with plasmas hoping to receive his support. Helmut Schwarz from the institute was willing to join in the organization of this conference. His contact to Nicolaas Bloembergen the later Nobel Laureate - who was then a most prominent after dinner speaker at the first conference - proposed the name LASER INTERACTION AND RELATED PLASMA PHENOMENA which is now a trade mark for 30 years.

The need for such a conference was just obvious at the time. From my position at Westinghouse Research in Pittsburgh I had joined the very first Gordon Conference in the Mount Rainier area in September 1967 about this topic and my informal presentation of the nonlinear force in plasmas at laser interaction due to dielectric effects was well absorbed by Allan Haught from United Aircraft and reproduced then in his report. In 1966 also the first meeting with less than 10 persons was held in Frascati called "European Conference on Laser Interaction with Matter (ECLIM)". I had in mind that the conference should not be a regional one but a fully international one and though informal and with not necessarily finally polished results should be documented in proceedings for which Plenum Press in New York was interested.

Though the conference was aiming to look into all kinds of interactions, espe-

Dedicated to Professor Edward Teller to his 90th birthday on January 15, 1998

cially that of extremely high intensities opening basically new physical processes as will be elaborated in the following, the application for exothermic production of fusion energy for power stations was one of the very attractive goals. This topic was indeed close to highly classified work around the world and this is the reason why the conference began from the very beginning with a lot of complexities including the fact that no financial support was available from outside. Nevertheless the numerous persons working on this topic were most enthusiastic and whether working in open or in restricted areas and despite all polarization during the cold war, the conference offered a meeting point to look into the solution of inexhaustive, clean and low cost energy from nuclear fusion by inertial confinement (ICF) for the future of mankind.

FUSION ENERGY

The first conference in June 1969 had indeed prominent speakers on this topic, John Dawson and, sent by Nobel Laureate Basov from Moscow, Gleb Slkizkov. Basov had indeed opened the first publication (with O. Krokhin) in February 1963 in Paris and had reported at the same international Quantum Electronics conference series in 1968 that he with Kryukov, Senatski et al may have produced the first few 100 fusion neutrons above the background level with laser pulses of some 10 ps duration. The experts were skeptical alone in not knowing the detailed properties of the laser pulses. But it seems that these had the crucial property of very low prepulse levels. Just this technique was presented in June 1969 at the first conference by Francis F. Floux from the Limeil Laboratory near Paris when firing laser pulses on frozen deuterium. Sklizkov had to be escorted through some US-Laboratories and I recall that he asked Moshe Lubin in Rochester what technique he used for eliminating the prepulse. Moshe said his equipment is just back in the Workshop.

It is a tragedy that the must-not publishing Gordon conference in August 1969 had Lubin's presentation that he did measure 10^2 to 10^3 fusion neutrons to which James Tuck from Los Alamos commented that he can rub his trousers and will generate neutrons. Beginning of September 1969, Francis Floux presented a similar paper as before in June at the Laser Conference in Belfast nothing saying about neutrons; only when asked in the discussion he disclosed that he had about 10^4 neutrons. Fortunately he was ready to write an appendix to his paper for LIRPP1 with these facts which were measured indeed in June and before but not sufficiently clearly confirmed. Indeed the main trick was to suppress any prepulse by a factor of 10^6 of the intensity. It was noticed that both Sklizkov and Lubin in their discussions at and around LIRPP1 possibly had some views in this direction.

At the following ECLIM conference November 1969 in Paris, then visited by Basov and other prominent colleagues, waiting for in important rally on neutron counts, S. Witkowski began his talk "We in Garching have 10^9 fusion neutrons" after which the 20 persons congregation behaved like in Gogol's "Revisor" last act:

they looked petrified from this shock. Then Witkowski continued "but in our theta pinch".

The second conference LIRPP in June 1971 in the newly established Graduate Center in Hartford Conn., downtown, had Sklizkov's presentation of 10^4 and more very detailed analyzed DD fusion neutrons. Paul Harteck from RPI, who with Mark Oliphant and Lord Rutherford had performed their very first DD fusion reaction in 1933, reported about Rutherford's rule on safety: use a clean table, use not more than 1 mg tritium and do it fast. He asked Sklizkov what safety precautions they have in the Lebedev Institute in Moscow. Sklizkov said he would volunteer to stay next to the target chamber if 10^{14} neutrons would be produced, but regrettably they did not have so many. He showed also the design of the nine beam glass laser "Kalmar" which was then called the "Sklizkov monster". This all contributed to the decision in September 1972 at the International Quantum Electronics Conference in Montreal that Edward Teller, John Nuckolls and others disclosed the then partially declassified work form Livermore: computations with ablative compression of DT plasma to 10,000 times the solid state producing interesting fusion gains with specially tailored laser pulses in the kJ range.

This was the starting point for large scale experiments at Livermore, Rochester, and KMS Fusion as well as in Japan (C. Yamanaka). France had begun a large scale program before. The German Government - definitely after consultations with other partners - offered DM100Mill as start money for a project (Dr. Menden) but the scientists in charge at the Garching Max-Planck-Institute did not accept the money as it was reported to may witnesses by Richard Sigel.

LIRPP No.3, 1973 joint all known laboratories (with exception of the Chinese) where laser fusion neutrons were produced. John Nuckolls and Keith Brueckner presented their fascinating results of computations and experimental programs. The after-dinner speaker was Edward Teller (Fig.1) after he gave a dramatic press conference before at the office of the Rensselaer Polytechnic Institute in Troy, NY, which was the venue of the conference.

The number of fusion neutrons was then indeed growing from conference to conference up to 10^9. Special results form the Japanese activities were most appreciably presented for the first time at these conferences. At LIRPP5, C. Yamanaka reported the first published result that laser irradiation of uranium produced isomeric states in ^{235}U by the then theoretically elaborated NEET (nuclear excitation by electronic transitions) as measured from the well known gamma emission of the excited uranium nuclides. Another hit from Japan was the sensational first presentation at LIRPP 1985 by C. Yamanaka and S. Nakai: 10^{12} neutrons from DT gas filled micro balloons with very thin glass shell. The empirically discovered secret was that one had to use a stagnation-free adiabatic compression. Non-adiabatic shock wave generation produced low gains and very nonuniform x-ray emission. The high gains followed the adiabatic self similarity volume compression and after the increase to 10^{13} neutrons (LIRPP 1989) and after some encouraging computations in Japan following the adiabatic model towards a volume ignition, this direction was given up later in favor of the shock producing spark ignition schemes and no higher neutron gains were reported from this side.

Fig. 1 At the press conference at LIRPP3,1973,at Rensselaer Polytechnic Institute,Troy,NY,from the right:George Baldwin (local organizer), Edward Teller, Helmut Schwarz, and Heinrich Hora

The next step for neutron production followed the introduction of the Omega upgrade laser as reported to be operational in April 1995 when LIRPP presented the Edward Teller Medal to R. McCrory of ILE Rochester. At LIRPP 1997, W. Seka et al confirmed that they produced $2x10^{14}$ DT fusion neutrons after 4 weeks before at an IAEA meeting in Osaka only half of this value was confirmed. The laser pulses of 30kJ with 7% hydroefficiency *produced then core gains G = 30%* (G=fusion energy per laser energy in the compressed plasma). This is more than ever has been produced by April 1997 even with magnetic confinement fusion though the latter was funded by about ten times higher capital. To the use of the core gain for this comparison, Mike Campbell said at LIRPP 1989 "Even Harold Furth from Princeton had to admit that this definition has to be accepted".

ADIABATIC SELF SIMILARITY COMPRESSION
AND IGNITION

As recently noticed (see contribution by the author with Azechi et al of this volume), the just mentioned record fusion neutrons of Rochester could be explained in a very straight forward way by adiabatic compression. It was rather a surprise that an alternative behavior of the long years measured decrease of maximum compression density in laser produced plasmas at increasing neutron gain (see Fig. 13.15 [1]) was expressed by an increasing quadratic law at isothermal conditions. This line - derived from the Osaka experiments of 1988 [2] with the then record gains - exactly agreed with the Rochester measurements of 1995 (10^{14} neutrons at compression density of $4n_s$ where n_s is the solid state density). We could then show in a very convincing and easily understandable way, how the fusion core gains calculations at optimized conditions [1] fully reproduced the results of the best gains in Rochester, Livermore, Osaka and Arzamas-16 with respect to density, gain and temperature as measured. A strong disagreement was seen when using an earlier fast pusher experiment understanding that then the strong shock wave creation was very far away from the ideal adiabatic conditions.

The adiabatic compression is indeed just the contrary what is the fashionable aim in laser fusion: the spark ignition. Using either the *isochoric* [3] or the *isobaric* [4] scheme of compression, a hot center in the compressed DT plasma is the aim surrounded by high density low temperature outer plasma and a two dimensional fusion detonation wave should be initiated burning the outer plasma. This spark ignition - whether the plasma is compressed by direct drive or by indirect drive via hohlraum radiation produced in a laser or beam irradiated capsule - is "possibly the most difficult point in ICF..." [5] because it is sensitive against Rayleigh-Taylor instabilities and needs a very high uniformity in the ignition sphere: if it ignites in one direction and not in the other direction, all is lost.

These difficulties can be overcome if a much more "robust" volume ignition scheme [6] is used. This is based on an *ideal adiabatic* compression as used by Dawson [7] and mathematically clarified in LIRPP 1 (p. 365) which obviously was realized in the experiments of Osaka [2] and Rochester [8] when the highest gains were the aim and not any shocks for a central spark. The calculations for the simple fusion burn agreed with the above mentioned experiments ideally. However these gains were far too low for conditions of a fusion reactor and the core gain formula (LIRPP 2, p. 386) is algebraically identical with Kidder's n_oR-value, Eq. 13.8 of Ref., [1] (n_o is the DT density at highest compression with a plasma radius R). In both calculations, that of the author and that of Kidder, no alpha reheat and no partial reabsorption of bremsstrahlung was included and the gains were indeed very low even at high compression and Nuckolls et al and Brueckner et al, LIRPP 3, introduced the spark ignition for very high fusion gains.However, after including alpha self heat and bremsstrahlung reabsorption into the adiabatic scheme, a volume ignition in the whole pellet like in a diesel engine without two dimensional combustion waves from a spark was generated [9]. Mostly the alpha heat acted as a kind of an additional strong driver energy and could heat up the ions to 200 keV resulting in such high gains, that similar values as at spark ignition were reached.

This volume ignition was first reproduced by John Wheeler with Ron Kirkpatrick three years later, then by Takabe, Mima and Nakai, by Tahir and Hoffmann, by Martinez-Val et al, by He et al, and even by Meyer-ter-Vehn with Anisimov or Aoki, and by Atzeni who initially were strongly fighting the volume ignition but honestly in their research finally calculated the high volume ignition gains of similar values as spark ignition [10].

This conference series was showing from the very first volume (p. 273) the continuous development of this scheme including contributions in this volume by several authors and including the first numerical observation of the ignition in the 1976 conference (LIRPP 4: Ray et al). Since this is not all believed yet, the computer code has been published (LIRPP 10: Stening et al) such that everybody can check the results though they were reproduced by many other authors [10].

Using this code, one can easily reproduce the result of a core gain of 400 (spark ignition is assumed to be about two times better) with 10kJ input energy if compression of $10,000n_s$ is used. This confirms very closely the results of John Nuckolls 1972 which was later criticized strongly as being too optimistic. Using the fully open code (LIRPP 10: Stening et al) one can easily prove under what conditions Nuckolls was rather right and not wrong!

The first application of ignition was devoted to find conditions for the ideal fusion reaction of protons with Boron(11). Thanks to the self heat at volume ignition, some first positive results were presented in 1976 at LIRPP4 (Ray et al) in order to follow the main idea of Miley to have clean fusion with no neutron generation. At that time Ray Kidder was saying at LIRPP 4 that the hydrogen-boron fusion is not for the year 2000 but for the 4000. This reaction is so clean that it is generation of nuclear energy with less radioactivity than burning coal! However it turned out [1], (LIRPP 10: Stening et al) that one needs drivers of higher efficiency than 30% with 100 MJ pulses and compression to 100,000 times the solid state density. This is well possible within 50 years from now, the same time when a tokamak power station is expected [11], if the less clean DT power station (at least working with volume ignition and whatever drive) will be available within 15 years in a crash program and some years later with the higher compression than the today measured 1000 times the solid state [12] may produce energy three times below the today's lowest cost of energy (LIRPP 12: Höpfl et al).

More sophisticated and more relaxed results for hydrogen boron may be derived from more sophisticated analyses than just mentioned as e.g. presented by Martinez-Val et al in this present volume.

LASER DESIGN

The topics of this conference series is closely involved with the achievements in laser technology. Though there are now special topics conferences on high power laser development, LIRPP has always had to include reports about the main achievements or proposed developments and had to put this at the beginning of the earlier proceedings.

There was concentration on solid state lasers, first on ruby, then on the bigger growing neodymium glass lasers. Skilzkov's monster (Kalmar) was mentioned before, then came his Delfin laser where scientifically unique measurements were achieved with unique diagnostics (shocks in low pressure gas, Förster's x-ray intense enlargement of x-ray pictures using bent crystals, gracing incidence x-ray spectroscopy etc., apart from the usual diagnostics) which were in many cases firsts reported at the conference Series. Parallel to the big size lasers at Rochester (Omega) or at Osaka (Gekko 12) the big science at Livermore was pushed by John Emmett culminating in the Shiva and the Nova laser, all reviewed at LIRPP conferences.

If the new Omega design (LIRPP 12: McCrory) is reaching the specifications of Nova now even for indirect drive experiments, one has to realize that the design of Nova was so many years earlier and again the design of Omega is being overtaken again in costs and size by the beamlet laser for NIF (national ignition facility) at Livermore in collaboration with Rochester, Sandia and Los Alamos which has the capacity of Nova with a beam cross section of 80cmx80cm. Proudly we can mention that this cassette type glass laser was presented at LIRPP 1985 by Manes et al from Livermore. There were indeed several improvements since with multi pass operation etc., but it is remarkable that Livermore could well have built a NIF laser nearly ten years earlier than it is now being financed. If an energy crisis with catastrophic consequences would happen in the next dozens of years because inertial fusion power stations are not available, one should remember that the mentioned (unnecessary) ten years loss may cost then trillions of dollars damage.

In addition to the glass lasers the developments of carbon dioxide lasers at Osaka (Lekko) and Los Alamos (Helios, Antares) were covered by LIRPP as well as Reed Jenssen's (Los Alamos) HF laser and the KrF excimer laser development. The most significant development of the iodine laser by K. Hohla (Garching) was included from its first success in LIRPP which development was to some extent continued there but was taken up on a very large scale with ISKRA-5 at the Arzamas-16 institute in Russia[13].

Nuclear pumped lasers as pioneered by George Miley et al and other teams in the USA and later also in reports from Russia, was a continuous topic of the LIRPP conferences. The scheme elaborated by Miley for 50MJ μs laser pulses at LIRPP 1979 was one special highlight not only for laser fusion but also for the then appearing SDI development of powerful lasers. LIRPP10 had the first report of a visible (sodium) nuclear reactor pumped laser.

Of the more exotic laser schemes, the gamma ray laser was discussed since early conferences and very competently continued by Carl Collins (U. Texas). The advent of the laser-plasma pumped x-ray laser was from its beginning at LIRPP mostly from the Livermore teams, but also form France (Limeil), England (Rutherford Lab), Japan (Tokyo Univ.), China (Shanghai), and other places. The failure of Hagelstein's NOVETTE ps pumped x-ray laser was evident since the calculations had classical and not anomalous absorption included. The success with longer pulses was rather an accident (Maxon, Livermore) as discussed at LIRPP or concluded from plasma phenomena (Brunner and John, Berlin-Adlershof).

The greatest highlight from the last years development are the femtosecond lasers which were presented 1993 at LIRPP11 by Mourou with the remarkable concept of a strength of alexandrite material for optics and diffraction mirrors well reaching into the range of Exawatt lasers. With the more realistic facts of available materials, Mike Perry et al from Livermore presented their work with really 2 Petawatt laser pulses which were scheduled to work just the week after the present conference. Nevertheless J.A. Koch of Livermore has a 100 TW laser since some time at operation and a most interesting 30 fs 10^{18} W/cm^2 laser is working at Pisa (these proceedings: L.A. Gizzi et al, Pisa, Italy).

THE STORY OF ION ACCELERATION AND LASER ACCELERATION OF ELECTRONS

The unusual physics and the new stage nonlinear phenomena (though mostly covered by simple classical physics but with more ugliness than known from plasma theory) were seen from the very beginning from the ions generated in laser produced plasmas. This topic of physics accompanied nearly all the conferences.

It was indeed strange. With the spiking laser pulses up to 1963, irradiation of targets with powers less than a MW resulted in heating of the targets and the plasma surface fully following the classical thermodynamics and plasma hydrodynamics with temperatures of few eV and with emission of ions of few eV energy and electron emission following the Languor-Child law. When Linlor fired 10 MJ Q-switched pulses he measured up to 10 keV highly charges ions and Honig found electron emission with more than 100 times higher electron densities than permitted by the classical space charge theories. I am proud that we had these two pioneers speaking at the first LIRPP conference. The ion energy increased quadratically on the laser power as it was permitted to be published in Canada (Isenor) but the US journals rejected all submissions. Therefore it was the correction to the truth when we were able by LIRPP1 to publish the long years waiting similar experiments of Helmut Schwarz.

When A.W. Ehler, LANL,just published his MeV ion energies, similar MeV ions resulted with neodymium glass lasers (B. Luther-Davies et al) whose paper was rejected by the Quantum Electronic conference in Amsterdam. Fortunately Optics Communications had accepted the publication. The long march to understand this all by ponderomotive and (since 1974/5) by relativistic self focusing [14] was again reflected at the LIRPP conferences. One of the pioneers, M.S. Sodha, contributed to the discussions. When in 1982 at a conference on laser accelerators, Chan Joshi said that the quarter ot half GeV ions followed my prediction from the relativistic self focusing, A.W. Ehler was then told that his results may be classified. A Los Alamos report for one of the Anomalous Absorption Conferences "not to be published" is the only documentation about the 500 MeV ions, again following relativistic self focusing [15].

Since there are three groups, the very fast from relativistic self focusing, a thermal one and one without separation by ion charge number and well suggested as due to hot electrons, could now be quantitatively explained [15] combining quiver motion with (quantum-) collisional thermalization. Recent results with iodine lasers producing 52 times ionized tantalum were one of the topics reported in these proceedings.

Acceleration of *electrons* was indeed in LIRPP1 with the phenomenon reported by Honig about the 1000 time higher electron emission currents than classical space charge theory permitted. This was confirmed later even to higher numbers by Siller et al (LIRPP 3). Then came the Boreham experiment (see Luther-Davies et al LIRPP5 and 6) with a non-Liouvillian emission of keV electrons radially from the laser beam and with the paper by Viera et al, LIRPP 6, p. 203, predicting the amount of the forward direction of the electrons due to conservation of the momentum of the nonlinearly transferred optical energy what was exactly measured only recently by D. Meyerhofer et al.. These problems will be discussed more in the following section.

The aim to use the very high electric field amplitudes in a laser focus to accelerate electrons is a long dream though the field goes into the wrong direction. Again this topic was in the focus of the conferences at times where controversial views were still frustrating progress. Now the free wave acceleration (J. Woodward and A. Kerman, Livermore) appears what has been presented at LIRPP since 1989 by the group form Giessen/Germany. This acceleration of electrons by lasers in vacuum without plasma effects has been shown to be involved in pioneering experiments of R. Umstadter et al and other places and is in focus of the forthcoming 2 PW experiments at Livermore. The acceleration effects with plasmas due to the very high amplitude longitudinal (Langmuir) oscillation fields driven by lasers as seen from the genuine two fluid plasma model and the beat wave acceleration indeed may provide additional effects. It was at the LIRPP conference 1989 that a relief of clarification came in the tutorial of Chan Joshi, explaining that the beat wave acceleration is a half (Langmuir-) wave length process on which basis all is then very easily understood. A competitive acceleration process due to the very large amplitude very arbitrary plasma oscillations driven by the laser field, as seen first from the genuine two fluid computations, explained double layer effects and internal electric fields which well could accelerate electrons to the considered very high MeV energies (see Eliezer, Szichman et al in LIRPP7 to 10).

The problem remains now what the stopping length of the relativistic electrons will be for the purposes of the fast ignitor. This scheme [16] would arrive at same fusion gains as with the classical ablation-compression-heating by ns pulses but with 10 times higher efficiency since the energy deposition should be performed with picosecond laser pulses depositing its energy with low losses in the center of the precompressed moderately heated plasma. Just the conference with the reports in the present proceedings indicate the problems about stopping length of the electrons and ions which receive the laser energy after relativistic self focusing [14] of the laser beam and squeezing of beamlets by magnetic fields [17]. This all merges the long years - and LIRPP characteristic - topic on energetic ions and electrons with the now fusion relevant fast ignitor.

THEORY OF FORCES AND INSTABILITIES

The last 35 years pioneering stage of laser-plasma interaction is much more determined by experiments mostly disclosing unexpected new properties, and was less guided by theory which mostly had difficulties to keep up with explanations of the observations. Nevertheless basic quantum physics and relativity was developed (LIRPP10, p. 11) and there were few theoretical aspects which were virtuously developed and basically clarified at least as one important subsection of the whole field. Indeed there was the work initiated by Erich Weibel 1957 at TRW and nearly simultaneously elaborated by Boot et al in England and Gapunov and Miller in Russia showing that the high frequency fields of standing waves produced forces by time averaged gradients of the electric field square, which value was formally identical with the ponderomotive force known form electrostatics. The step from the HF-conditions for electrons without plasma in vacuum interacting with laser fields to the case of *dielectric plasma properties* was done *for the first time* resulting in the *nonlinear forces* f_{NL} in LIRPP1, p. 341. From then on this topic of nonlinear forces was discussed in most of the conferences.

The unique derivation of one of the (initially well derived) term by John Stamper form NRL (LIRPP 4, p. 721) found special attention and was called Stamper term. The complications for obliquely incident laser radiation on an inhomogeneous plasma led to a necessary extension of Schlüter's initial two-fluid plasma theory by the necessary introduction of new nonlinear terms without which no conservation of momentum was possible. This result was restricted, however, to non transient conditions only.

A long years controversy about the correct description of the transient case was indeed outside LIRPP until Zeidler Mulser and Schnabel clarified that the six different theories could be merged at least approximately. LIRPP7 p.347 came into action then with an additional term to be added to the last mentioned result arriving only by this way in the finally complete transient nonlinear force f_{NL} (LIRPP7, p. 347) as later confirmed by Rowlands that this and only this formulation is Lorentz and gauge invariant.

Another important property in laser produced plasmas are the *instabilities*. Most significant are the stimulated Raman, the stimulated Brillouin, the two-stream instability and several more. These instabilities were indeed first developed from the interaction of intense microwaves with plasmas and the first work can be traced back to V.N. Oraevsli and R.Z. Sagdeev (1963). The following work by DuBois and Goldman (see LIRPP4), Silin and Nishikawa were all reduced to a fully unified description by the ingenious work by F.F. Chen he presented for the first time at LIRPP 3, p. 291: This all could be *described in a common and unified* way by the nonlinear force f_{NL}.

Just this work opened the question whether and how to describe the forces by macroscopic field quantities like *gradients of the Maxwellian stress tensor* or by single *electron motion* driven by the Lorentz force including dielectric effects of the plasma. This arrived at the *Chen-paradox* which was clarified at LIRPP6, p.439 but the problem appeared again at several interpretations. In the Boreham experiment [18] with radial emission of keV electrons from a 10^{16} W/cm^2 neodymium glass laser beam in low density helium, the ponderomotive potential explained how half of the maximum oscillation energy of the electrons was converted into translative electron energy independent of polarization. When calculating the single electron motion with the transverse laser fields, a 100% polarization dependence appeared contrary to observation. What was the problem? Maxwell's equations for a laser beam lead also to longitudinal (!) laser field components and only with these as exact solutions the result is polarization independent [19]. This teaches that one has to work *very precisely and without approximations in nonlinear physics* otherwise trouble like chaos etc. appears.

An enormous amount of experimental and theoretical work was devoted to the *instabilities* which well exist as measured by the 3/2 harmonics emission from laser produced plasmas. Sometimes very special density plateaus had to be generated artificially to produce the stimulated Brillouin instability. LIRPP did not follow up all the details of these developments which were more discussed at the Anomalous Absorption Conferences. Even if not the last word has been spoken about these instabilities, the development went over to demonstrate that the smoothing of the laser beams for a most uniform interaction of the laser beams with the plasma resulted in a strong reduction of these instabilities as pioneered especially by the NLR but also by theories of Dragila parallel to experiments by Luther-Davies et al.

Higher harmonics can well be produced by other phenomena. The measured wide spread second harmonics emission of the extented inhomogeneous plasma corona of nearly constant intensity (never explained by instabilities) could be understood by double layer effects and semianalytical analysis (LIRPP8,p. 239). A further phenomenon is the 5 to 20 ps pulsating interaction as discovered by Luther-Davies et al and explained in a straightforward way by computations with the genuine two-fluid model (M. Aydin et al, LIRPP 10, p. 181). Indeed some indications of this stuttering interaction was derived numerically at Rochester 1974 and seen in experiments of Lubin et al but the crucial significance and how this stuttering interaction can be suppressed and a low reflectivity reached for a laser radiation deposition in the deep corona of the plasma became evident only after clarifying the stochastic pulsation problems during the last few years. This is essential for direct drive and was a crucial obstacle against laser fusion before.

PAIR PRODUCTION

The LIRPP conferences can be especially satisfied that their workshop-like informal presentation permitted to discuss some very vague or futural aspects. One topic of this kind is the pair production by very intense laser intensities. The first indications were reported by Shearer et al (LIRPP3) where intense x-rays were observed from laser produced plasmas after 10 cm lead screens. It was just the type of the conference that the paper by Shearer et al in Phys. Rev. said nothing about electron positron pairs due to a trident mechanism but the formulations at LIRPP permitted an estimation and several details to claim the production of the pairs.

Theoretical estimations about pair production by several different mechanisms (trident, or one electron in an intense laser field, or pairs from the vacuum by vacuum polarization (Euler Heisenberg)) were discussed at the early LIRPP conferences and very futuristic scenario later on how to produce anti-hydrogen. At least the trident-laser mechanisms permitted anti-particle production by a large number of magnitude more efficient than the pair production in accelerators because of the very high particle density in the laser focus.

The more it was important to see e.g. the paper by T. Kotseroglou at LIRPP11, where a very sophisticated experiment was established on the basis of a concept of D. Reiss in combination of a number of specialized teams from different centers, to look into the interaction of the 30 GeV electron beams from SLAC with 10^{18} W/cm^2 laser intensities. This intensity is at least by a factor 10 below andytrident pair production but the theory concluded other interaction mechanisms. While at LIRPP11 only the fascinating steps into these experiments were described and - understandably - not recorded in the proceedings, LIRPP13 has now the results of the admirable success: David D. Meyerhofer reported the results of the positron production at SLAC as a result of the long years work.

RELATED PLASMA PHENOMENA

It was most visionary by Nicholaas Bloembergen that he proposed the title of the conference 30 years ago: it is not only the most fascinating interaction of the lasers but there are alos a lot of neighbor problems with the plasmas. One has to begin with the related directions to arrive at inertial confinement fusion energy. One of the initial pioneers, G. Yonas, was at the early conference (LIRPP3) for orientation of his work first going into electron beam fusion, then pioneering the light ion beam fusion at Sandia before going to be in charge of the similarly related organization of SDI. His views were always cooperative for the laser interaction even if he was pioneering the alternative line.

I cannot suppress my satisfaction that the nonlinear physics developed by LIRPP especially came to the explanation what cardinal mistake was done in 1951 when fusion energy was directed towards magnetic confinement (see LIRPP6, p. 929). When E.O. Lawrence, Sir Mark Oliphant and other giants liked to harvest fusion energy they were thinking of ion beams. But then came Lyman Spitzer and explained that the cross section of fusion interaction for 100kV deuterons with tritons is more than 300 times less than the interaction with electrons. While one does produce fusion reactions as in the initial fusion experiment with Rutherford 1933, the heat sink of cold electrons prevents any exothermic reaction and one has to heat electrons and ions to several 10Million degrees (confined by magnetic fields) such that the electron energy losses do not count. Beam fusion must never work, was the conclusion. This argument is correct by logic, mathematics and physics. But it is nevertheless wrong: it is linear while the today's laser or beam fusion is basically nonlinear.

Since the time when George Miley followed the late Helmut Schwarz as co-director from LIRPP6 on, particle beam fusion was represented with key contributions. This was just the time, the big heavy ion beam fusion and the electron beam and then the light ion beam fusion projects were opened. LIRPP was on the right way, since the ICF work with NIF is now integrating all national laser and particle beam activities.

Another important related field was the ICF fusion reactor technology. While LIRPP could never be a topic conference on each of these related fields, it was imperative to keep in touch by key reports. Just the reactor technology indicated the advantage of ICF over magnetic confinement. In the tokamaks, the wall erosion is a mortal problem. Some materials have an erosion of meters per year, only tungsten using "plasma sweeping" arrives at 5cm per year "and this is too much" [20]. Maisonier, the director of fusion research at the EU in Brussels [11] expects the magnetic confinement reactor not before 50 years where the question is open whether it will then be economically competitive. In ICF, the wall erosion is not at all a problem. The reactor has a 50cm blanket with liquid lithium salt or lithium ceramic pellets which usefully absorb and convert most of the micro-explosion reaction products before reaching the reactor wall. Any explosion shock is reduced by gas dynamics and due to the much higher nuclear energy compared with chemical explosives. The whole concept of the tokamak came into a critical light since all the predicted goals of the very big experiments were never reached as could be explained now by a turbulence mechanism [21]. On the other hand the worries about ICF with respect to Rayleigh-Taylor instabilities were mostly overcome by the experimental achievements: the measurement of more than 1000times the solid state in laser compressed polyethylene [12] and the smooth surface of the laser produced plasma indicated that the instabilities are less important. One reason may be the stabilization by surface waves [22].

LIRPP also kept in touch with other related plasma phenomena, as material treatment (though taking only significant presentations compared with the big topical conferences). Laser evaporation of barium titanate without chemical change was a topic in LIRPP1 up to the laser ion source for accelerators and other applications in the present proceedings.

FINAL REMARKS

One may ask at this stage whether the topic LASER INTERACTION AND RELATED PLASMA PHENOMENA has now been sufficiently accompanied and documented during the last 30 years and whether this series can fulfill its importance further on. There are many similar places of conferences. The international Quantum Electronic Conference (the last in July 1996 in Sydney) was indeed covering most important early highlights of our topics, as 1963 the first publication on laser fusion by Basov and Krokhin, 1968 the first (weak) laser fusion neutrons (Basov et al), 1970 the first ponderomotive self focusing theory in plasmas and the isentropic fusion gain formula (Hora and Pfirsch), and 1972 the most sensational disclosure of the Livermore laser fusion work by Edward Teller, John Nuckolls et al. But now, this conference is devoted to medium laser intensities, chemistry, nothing on laser fusion or on relativistic interaction effects. The "Anomalous Absorption Conference" is a continuation of the earlier closed Gordon conferences with still main participation form the USA and with topics on instabilities which are in some cases obsolete now and where significant new and crucial phenomena as the 5ps stochastic pulsation were overlooked or marginally covered only. The European ECLIM conferences are indeed more international. But as we see from the Teller-Lecture of G. Velarde in these proceedings, the ICF is highly suffering from the domination of the magnetic confinement establishment there. The attempt by M. Key and E. Fabre few years ago who brought together all European science ministers whom they persuaded such that they were excited and determined to start a big ICF activity, this all failed when the ministers came to their offices and were knocked down by the magnetic fusion lobby there. The following attempt in 1996 where the essential committees requested 10% of the fusion budget for ICF was finally reduced to 0.5%. ECLIM joins the European enthusiasts who like to devote their excellent experience to ICF but cannot move much.

Therefor the only places where ICF is moving on a large scale are the USA (now with ideally united activities around NIF) and Japan. The international participation is formalized and the contributions of smaller worldwide activities to the main stream with significant individual contributions are to be somehow merged. This was done to a good level in the past by LIRPP and may be increased in the one or other way by the new Director(s) in the future. The commitment - especially for Livermore - consists in the celebration of Edward Teller and his incommensurable merits and contributions to this field. Thanks to E.M. Campbell, J.L. Nuckolls and W.J. Hogan during 1995/6 with the long years operations of G.H. Miley, the formal connection of the Edward Teller medal with the American Nuclear Society and the envisaged endowment fund is a further new development which is most promising.

[1]Hora,H.*Plasmas at High Temperature and Density*(Springer,Heidelberg 1991)

[2]Takabe, H., Mima, K., et al., Phys. Fluids **31**, 2884 (1988)

[3]Kidder,R.,Nucl.Fusion,**16**,406(1976); Bodner,S.,J.Fusion Energy,1,221(1981)

[4]Meyer-ter-Vehn, J., Nuclear Fusion **22**, 561 (1982)

[5]Meyer-ter-Vehn, J., *Plasma Physics and Controlled Nuclear Fuison Research* 1994 (IAEA, Vienna 1996), Summmary on ICF

[6]Lackner, K. S., Colgate, S.A., et al, *Laser Interaction and Related plasma Phenomena*, G.H. Miley, ed., AIP Conf. Proceedings No. 318 (Am.Inst.Phys., New York, 1984) p. 346

[7]Dawson, J.M., Phys. Fluids **7**, 981 (1964); Hora, H., Inst. Plasma Physik Garching, Rept. 6/23; Hora. H., and Pfirsch, D., *6th Intern. Quantum Electr. Conf., Kyoto, Sept. 1970, Conf. Digest*, p. 10

[8]Soures, J.M., McCrory, R.L. et al, Phys. Plasmas **3**, 2108 (1996)

[9]Hora, H., and Ray, P.S., Zeitschr. f. Naturforschung **33A**, 890 (1978)

[10] Kirkpatrick, R.S. and Wheeler, J.A., Nuclear Fusion **21**, 389 (1981); Mima, K., Takabe, H., and Nakai, S., Laser and Particle Beams **7**, 467 (1989); He, X.T, and Li, Y.S., *Laser Interaction and Related Plasma Phenomena*, G.H. Miley ed. AIP Conf. Proceed. No. 318 (Am. Inst. Phys. New York 1994) p. 334; Tahir, N.A., and Hoffmann, D.H.H., Fusion Engineering and Design, **24**, 418 (1994); Martinez-Val J.-M., et al Laser and Paritcle Beams **12**, 681 (1994); Anisimov, A. Oparin, A. and Meyer-ter-Vehn, J., *GSI Annual Report 1994*, p. 44; Atzeni, S., Jap. J. Appl. Phys., **34**, 1980 (1995)

[11] Maisonier, Europhys. News, **25**, 167 (1994)

[12] Azechi, H., et al, Laser and Particle Beams **9**, 167 (1991)

[13] Kirillov, G.A., et al., Laser and Particle Beams **8**, 827 (1990)

[14] Hora, H., J. Opt. Soc. Am. **65**, 882 (1975)

[15] Haseroth, H., et al, Laser and Particle Beams, **14**, 393 (1996)

[16] Tabak, M., et al Phys. Plasmas, **1**, 1626 (1994)

[17] Pukhov, A., and Meyer-ter-Vehn, J., Phys. Rev. Letters, **76**, 3975 (1996)

[18] Boreham, B.W., and Hora, H., Phys. Rev. Letters, **42**, 776 (1979)

[19] Cicchitelli et al, Phys. Rev. **A41**, 3727 (1990)

[20] Rebout, P., Colloquium CERN, May 1991

[21] Dorland, W., and Kotschenreuter, M., Bull Am. Phys. Soc. **41**, Oct. (1996); Glanz, J., Science, **274**, (6.Dec.) 1600 (1996)

[22] Eliezer, S. et al., Physics Report **172**, 339 (1989)

AUTHOR INDEX

SUBJECT INDEX

Acknowledgment

This book would not have been created without the continuous attention and support by the late Edward Teller for which very special thanks are expressed. Thanks are also due to all the Board Members of Laser Interaction and Related plasma Phenomena (LIRPP) No. 10 to No. 13 and to that of Inertial Fusion Science and Applications (IFSA) 1999 to 2003, and also due to the Teller Awardees in the past being all involved in the sophisticated election procedures form a large number of very outstanding candidates under the discretion of ballots organized by the chairmen of the Awards Committee [George Miley (1991), Heinrich Hora (1993, 1995), John Nuckolls (1997) and Mike Campbell with Bill Hogan for the IFSA (1999, 2001, 2003)].

Since we were strictly reproducing the lectures from the preceding publications, the award for the Edward Teller Medal to Larry Foreman (Los Alamos National Laboratory) at IFSA1999 was honoured there but no lecture was delivered because of the his severe illness. We acknowledge that Mike Campbell as co-director of IFSA handed over the medal before Larry Foreman has passed away.

Thanks are due to giving permission of copyright by

- Plenum Press, New York for parts of Laser Interaction and Related Plasma Phenomena Vol 3 and Vol. 10
- American Institute of Physics for parts from the volumes of the AIP Conference Proceedings Laser Interaction and Related Plasma Phenomena No. 318 No. 369 and No. 406, where the contacts with the Editor-in-Chief of these series, Charles Doering has to especially acknowledged
- Elsevier Publishers for the use of parts of Inertial Fusion Science and Applications 1999 and 2001
- To Larry Suter and Hideaki Tanaka for sharing the copyright with the American Nuclear Society when preparing the proceedings for Inertial Fusion Science and Applications 2003.

H.H. and G.H.M.